Digital Image
Sequence Processing,
Compression, and Analysis

Computer Engineering Series

Series Editor: Vojin Oklobdzija

Low-Power Electronics Design
Edited by Christian Piguet

*Digital Image Sequence Processing,
Compression, and Analysis*
Edited by Todd R. Reed

*Coding and Signal Processing for
Magnetic Recording Systems*
Edited by Bane Vasic and Erozan Kurtas

Digital Image Sequence Processing, Compression, and Analysis

EDITED BY
Todd R. Reed

University of Hawaii at Manoa
Honolulu, HI

CRC PRESS

Boca Raton London New York Washington, D.C.

Library of Congress Cataloging-in-Publication Data

Digital image sequence processing, compression, and analysis / edited by Todd R. Reed.
    p. cm.
    Includes bibliographical references and index.
    ISBN 0-8493-1526-3 (alk. paper)
    1. Image processing—Digital techniques. 2. Digital video. I. Reed, Todd Randall.

TA1637.D536 2004
621.36′7—dc22   2004045491

Visit the CRC Press Web site at www.crcpress.com

To my wife, Nancy.

Preface

Digital image sequences (including digital video) are an increasingly common and important component in technical applications, ranging from medical imaging and multimedia communications to autonomous vehicle navigation. They are ubiquitous in the consumer domain, due to the immense popularity of DVD video and the introduction of digital television.

Despite the fact that this form of visual representation has become commonplace, research involving digital image sequence remains extremely active. The advent of increasingly economical sequence acquisition, storage, and display devices, together with the widespread availability of inexpensive computing power, opens new areas of investigation on an almost daily basis.

The purpose of this work is to provide an overview of the current state of the art, as viewed by the leading researchers in the field. In addition to being an invaluable resource for those conducting or planning research in this area, this book conveys a unified view of potential directions for industrial development.

About the Editor

Todd R. Reed received his B.S., M.S., and Ph.D. degrees in electrical engineering from the University of Minnesota in 1977, 1986, and 1988, respectively.

From 1977 to 1983, Dr. Reed worked as an electrical engineer at IBM (San Jose, California; Rochester, Minnesota; and Boulder, Colorado) and from 1984 to 1986 he was a senior design engineer for Astrocom Corporation, St. Paul, Minnesota. He served as a consultant to the MIT Lincoln Laboratory from 1986 to 1988. In 1988, he was a visiting assistant professor in the Department of Electrical Engineering, University of Minnesota. From 1989 to 1991, Dr. Reed acted as the head of the image sequence processing research group in the Signal Processing Laboratory, Department of Electrical Engineering, at the Swiss Federal Institute of Technology in Lausanne. From 1998 to 1999, he was a guest researcher in the Computer Vision Laboratory, Department of Electrical Engineering, Linköping University, Sweden. From 2000 to 2002, he worked as an adjunct professor in the Programming Environments Laboratory in the Department of Computer Science at Linköping. From 1991 to 2002, he served on the faculty of the Department of Electrical and Computer Engineering at the University of California, Davis. Dr. Reed is currently professor and chair of the Department of Electrical Engineering at the University of Hawaii, Manoa. His research interests include image sequence processing and coding, multidimensional digital signal processing, and computer vision.

Professor Reed is a senior member of the Institute of Electrical and Electronics Engineers (IEEE) and a member of the European Association for Signal Processing, the Association for Computing Machinery, the Society for Industrial and Applied Mathematics, Tau Beta Pi, and Eta Kappa Nu.

Contributors

Pedro M. Q. Aguiar
ISR—Institute for Systems and
  Robotics, IST—Instituto Superior
  Técnico
Lisboa, Portugal

Luis D. Alvarez
Department of Computer Science
  and A.I.
University of Granada
Granada, Spain

Guido Maria Cortelazzo
Department of Engineering
  Informatics
University of Padova
Padova, Italy

Thao Dang
Institut für Mess- und
  Regelungstechnik
Universität Karlsruhe
Karlsruhe, Germany

Edward J. Delp
School of Electrical Engineering
Purdue University
West Lafayette, Indiana, USA

Francesco G. B. De Natale
Dipartimento Informatica e
  Telecomunicazioni
Universita di Trento
Trento, Italy

Gaetano Giunta
Department of Applied Electronics
University of Rome Tre
Rome, Italy

Jan Horn
Institut für Mess- und
  Regelungstechnik
Universität Karlsruhe
Karlsruhe, Germany

Radu S. Jasinschi
Philips Research
Eindhoven, The Netherlands

Sören Kammel
Institut fur Mess- und
  Regelungstechnik
Universität Karlsruhe
Karlsruhe, Germany

Aggelos K. Katsaggelos
Department of Electrical and
  Computer Engineering
Northwestern University
Evanston, Illinois, USA

Anil Kokaram
Department of Electronic and
  Electrical Engineering
University of Dublin
Dublin, Ireland

Luca Lucchese
School of Engineering and
  Computer Science
Oregon State University
Corvallis, Oregon, USA

Rafael Molina
Department of Computer Science
  and A.I.
University of Granada
Granada, Spain

José M. F. Moura
Department of Electrical and
  Computer Engineering
Carnegie Mellon University
Pittsburgh, Pennsylvania, USA

Charnchai Pluempitiwiriyawej
Department of Electrical and
  Computer Engineering
Carnegie Mellon University
Pittsburgh, Pennsylvania, USA

Christoph Stiller
Institut für Mess- und
  Regelungstechnik
Universität Karlsruhe
Karlsruhe, Germany

Cuneyt M. Taskiran
School of Electrical Engineering
Purdue University
West Lafayette, Indiana, USA

Contents

chapter 1

Introduction

Todd R. Reed

The use of image sequences to depict motion dates back nearly two centuries. One of the earlier approaches to motion picture "display" was invented in 1834 by the mathematician William George Horner. Originally called the Daedaleum (after Daedalus, who was supposed to have made figures of men that seemed to move), it was later called the Zoetrope (literally "life turning") or the Wheel of Life. The Daedaleum works by presenting a series of images, one at a time, through slits in a circular drum as the drum is rotated.

Although this device is very simple, it illustrates some important concepts that also underlie modern image sequence displays:

1. The impression of motion is illusory. It is the result of a property of the visual system referred to as persistence of vision. An image is perceived to remain for a period of time after it has been removed from view. This illusion is the basis of all motion picture displays.
2. When the drum is rotated slowly, the images appear (as they are) a disjoint sequence of still images. As the speed of rotation increases and the images are displayed at a higher rate, a point is reached at which motion is perceived, even though the images appear to flicker.
3. Further increasing the speed of rotation, we reach a point at which flicker is no longer perceived (referred to as the critical fusion frequency).
4. Finally, the slits in the drum illustrate a vital aspect of this illusion. In order to perceive motion from a sequence of images, the stimulus the individual images represent must be removed for a period of time between each presentation. If not, the sequence of images simply merges into a blur. No motion is perceived.

The attempt to display image sequences substantially predates the ability to acquire them photographically. The first attempt to acquire a sequence of photographs from an object in motion is reputed to have been inspired by

0-8493-1526-3/2004/$0.00+$1.50
© 2004 by CRC Press LLC

a wager of Leland Stanford circa 1872. The wager involved whether or not, at any time in its gait, a trotting horse has all four feet off the ground.

The apparatus that eventually resulted, built on Stanford's estate in Palo Alto by Eadweard Muybridge, consisted of a linear array of cameras whose shutters are tripped in sequence as the subject passes each camera. This device was used in 1878 to capture the first photographically recorded (unposed) sequence. This is also the earliest known example of image sequence analysis.

Although effective, Muybridge's apparatus was not very portable. The first portable motion picture camera was designed by E. J. Marey in 1882. His "photographic gun" used dry plate technology to capture a series of 12 images in 1 second on a single disk. In that same year, Marey modified Muybridge's multicamera approach to use a single camera, repeatedly exposing a plate via a rotating disk shutter. This device was used for motion studies, utilizing white markers attached to key locations on a subject's anatomy (the hands, joints, feet, etc.). This basic approach is widely used today for motion capture in animation.

Although of substantial technical and scientific interest, motion pictures had little commercial promise until the invention of film by Hannibal Goodwin in 1887, and in 1889 by Henry W. Reichenbach for Eastman. This flexible transparent substrate provided both a convenient carrier for the photographic emulsion and a means for viewing (or projecting) the sequence. A great deal of activity ensued, including work sponsored by Thomas Edison and conducted by his assistant, W. K. L. Dickson.

By 1895, a camera/projector system embodying key aspects of current film standards (35-mm width, 24-frame-per-second frame rate) was developed by Louis Lumiére. This device was named the Cinématographe (hence the cinéma).

The standardization of analog video in the early 1950s (NTSC) and late 1960s (SECAM and PAL) made motion pictures ubiquitous, with televisions appearing in virtually every home in developed countries. Although these systems were used primarily for entertainment purposes, systems for technical applications such as motion analysis continued to be developed. Although not commercially successful, early attempts at video communication systems (e.g., by AT&T) also appeared during this time.

The advent of digital video standards in the 1990s (H.261, MPEG, and those that followed), together with extremely inexpensive computing and display platforms, has resulted in explosive growth in conventional (entertainment) applications, in video communications, and in evolving areas such as video interpretation and understanding.

In this book, we seek both to establish the current state of the art in the utilization of digital image sequences and to indicate promising future directions for this field.

The choice of representation used in a video-processing, compression, or analysis task is fundamental. The proper representation makes features of interest apparent, significantly facilitating operations that follow. An inappropriate representation obscures such features, adding significantly to complexity (both conceptual and computational). In "Content-Based Image Sequence Representation" by Aguiar, Jasinschi, Moura, and Pluempitiwiriyawej, video representations based on semantic content are examined. These representations promise to be very powerful, enabling model-based and object-based techniques in numerous applications. Examples include video compression, video editing, video indexing, and scene understanding.

Motion analysis has been a primary motivation from the earliest days of image sequence acquisition. More than 125 years later, the development of motion analysis techniques remains a vibrant research area. Numerous schools of thought can be identified. One useful classification is based on the domain in which the analysis is conducted.

In "The Computation of Motion" by Stiller, Kammel, Horn, and Dang, a survey and comparison of methods that could be classified as spatial domain techniques are presented. These methods can be further categorized as gradient-based, intensity-matching, and feature-matching algorithms. The relative strengths of some of these approaches are illustrated in representative real-world applications.

An alternative class of motion analysis techniques has been developed in the frequency (e.g., Fourier) domain. In addition to being analytically intriguing, these methods correlate well with visual motion perception models. They also have practical advantages, such as robustness in the presence of noise. In "Motion Analysis and Displacement Estimation in the Frequency Domain" by Lucchese and Cortelazzo, methods of this type are examined for planar rigid motion, planar affine motion, planar roto-translational displacements, and planar affine displacements.

Although there remain technical issues surrounding wireless video communications, economic considerations are of increasing importance. Quality of service assurance is a critical component in the cost-effective deployment of these systems. Customers should be guaranteed the quality of service for which they pay. In "Quality of Service Assessment in New Generation Wireless Video Communications," Giunta presents a discussion of quality-of-service assessment methods for Third Generation (3G) wireless video communications. A novel technique based on embedded video watermarks is introduced.

Wireless communications channels are extremely error-prone. While error-correcting codes can be used, they impose computational overhead on the sender and receiver and introduce redundancy into the transmitted bitstream. However, in applications such as consumer-grade video communications, error-free transmission of all video data may be unnecessary if the errors can be made unobtrusive. "Error Concealment in Digital Video" by De Natale provides a survey and critical analysis of current techniques for obscuring transmission errors in digital video.

With the increase in applications for digital media, the demand for content far exceeds production capabilities. This makes archived material, particularly motion picture film archives, increasingly valuable. Unfortunately, film is a very unstable means of archiving images, subject to a variety of modes of degradation. The artifacts encountered in archived film, and algorithms for correcting these artifacts, are discussed in "Image Sequence Restoration: A Wider Perspective" by Kokaram.

As digital video archives continue to grow, accessing these archives in an efficient manner has become a critical issue. Concise condensations of video material provide an effective means for browsing archives and may also be useful for promoting the use of particular material. Approaches to generating concise representations of video are examined in "Video Summarization" by Taskiran and Delp.

Technological developments in video display have advanced very rapidly, to the point that affordable high-definition displays are widely available. High definition program material, although produced at a growing rate, has not kept pace. Furthermore, archival video may be available only at a fixed (relatively low) resolution. In the final chapter of this book, "High-Resolution Images from a Sequence of Low-Resolution Observations," Alvarez, Molina, and Katsaggelos examine approaches to producing high-definition material from a low-definition source.

Bibliography

Gerald Mast. *A Short History of Movies*. The Bobbs-Merrill Company, Inc., New York, 1971.

Kenneth Macgowan. *Behind the Screen – The History and Techniques of the Motion Picture*. Delacorte Press, New York, 1965.

C.W. Ceram. *Archaeology of the Cinema*. Harcourt, Brace & World, Inc., New York, 1965.

John Wyver. *The Moving Image – An International History of Film, Television, and Video*. BFI Publishing, London, 1989.

chapter 2

Content-based image sequence representation

Pedro M. Q. Aguiar, Radu S. Jasinschi, José M. F. Moura, and Charnchai Pluempitiwiriyawej[1]

Contents

[1] The work of the first author was partially supported by the (Portuguese) Foundation for Science and Technology grant POSI/SRI/41561/2001. The work of the third and fourth authors was partially supported by ONR grant # N000 14-00-1-0593 and by NIH grants R01EB/AI-00318 and P41EB001977.

Abstract. In this chapter we overview methods that represent
video sequences in terms of their content. These methods differ
from those developed for MPEG/H.26X coding standards in that
sequences are described in terms of extended images instead of
collections of frames. We describe how these extended images,
e.g., mosaics, are generated by basically the same principle: the
incremental composition of visual photometric, geometric, and
multiview information into one or more extended images. Different outputs, e.g., from single 2-D mosaics to full 3-D mosaics, are

obtained depending on the quality and quantity of photometric, geometric, and multiview information. In particular, we detail a framework well suited to the representation of scenes with independently moving objects. We address the two following important cases: (i) the moving objects can be represented by 2-D silhouettes (generative video approach) or (ii) the camera motion is such that the moving objects must be described by their 3-D shape (recovered through rank 1 surface-based factorization). A basic preprocessing step in content-based image sequence representation is to extract and track the relevant background and foreground objects. This is achieved by 2-D shape segmentation for which there is a wealth of methods and approaches. The chapter includes a brief description of active contour methods for image segmentation.

2.1 Introduction

The processing, storage, and transmission of video sequences are now common features of many commercial and free products. In spite of the many advances in the representation of video sequences, especially with the advent and the development of the MPEG/H.26X video coding standards, there is still room for more compact video representations than currently used by these standards.

In this chapter we describe work developed in the last 20 years that addresses the problem of content-based video representation. This work can be seen as an evolution from standard computer vision, image processing, computer graphics, and coding theory toward a full 3-D representation of visual information. Major application domains using video sequences information include visually guided robotics, inspection, and surveillance; and visual rendering. In visually guided robotics, partial or full 3-D scene information is necessary, which requires the full reconstruction of 3-D information. On the other hand, inspection and surveillance robotics often requires only 2-D information. In visual rendering, the main goal is to display the video sequence in some device in the best visual quality manner. Common to all these applications is the issue of compact *representation* since full quality video requires an enormous amount of data, which makes its storage, processing, and transmission a difficult problem. We consider in this paper a hierarchy of content-based approaches: (i) generative video (GV) that generalizes 2-D mosaics; (ii) multilayered GV type representations; and (iii) full 3-D representation of objects.

The MPEG/H.26X standards use frame-based information. Frames are represented by their GOP structure (e.g., IPPPBPPPBPPPBPPP), and each frame is given by slices composed of macro-blocks that are made of typically 8×8 DCT blocks. In spite of many advances allowed by this representation, it falls short in terms of the level of details represented and compression rates. DCT blocks for spatial luminance/color coding and macro-blocks for motion coding provide the highest levels of details. However, they miss capturing pixel-level luminance/color/texture spatial variations and temporal (velocity) variations, thus leading to visual artifacts. The compression ratios achieved, e.g., 40:1, are still too low for effective use of MPEG/H.26X standards in multimedia applications for storage and communication purposes.

Content-based representations go beyond frame-based or pixel-based representations of sequences. Video content information is represented by objects that have to be segmented and represented. These objects can be based on 2-D information (e.g., faces, cars, or trees) or 3-D information (e.g., when faces, cars, or trees are represented in terms of their volumetric content). Just segmenting objects from individual video frames is not sufficient; these segmented objects have to be combined across the sequence to generate *extended* images for the same object. These extended images, which include mosaics, are an important element in the "next generation" systems for compact video representation. Extended images stand midway between frame-based video representations and full 3-D representations. With extended images, a more compact representation of videos is possible, which allows for their more efficient processing, storage, and transmission.

In this chapter we discuss work on extended images as a sequence of approaches that start with standard 2-D panoramas or mosaics, e.g., those used in astronomy for very far objects, to full 3-D mosaics used in visually guided robotics and augmented environments. In the evolution from standard single 2-D mosaics to full 3-D mosaics, more assumptions and information about the 3-D world are used. We present this historical and technical evolution as the development of the same basic concept, i.e., the incremental composition of photometric (luminance/color), shape (depth), and points of view (multiview) information from successive frames in a video sequence to generate one or more mosaics. As we make use of additional assumptions and information about the world, we obtain different types of extended images.

One such content-based video representation is called generative video (GV). In this representation, 2-D objects are segmented and compactly represented as, for example, coherent stacks of rectangles. These objects are then used to generate mosaics. GV mosaics are different from standard mosaics. GV mosaics include the static or slowly changing background mosaics, but they also include foreground moving objects, which we call figures. The GV video representation includes the following constructs: (i) layered mosaics, one for each foreground moving 2-D object or objects lying at the same depth level; and (ii) a set of operators that allow for the efficient synthesis of video sequences. Depending on the relative depth between different objects in the scene and the background, a single or a multilayered representation may be needed. We have shown that GV allows for a very compact video sequence representation, which enables a very efficient coding of videos with compression ratios in the range of 1000:1.

Often, layered representations are not sufficient to describe well the video sequence, for example, when the camera motion is such that the rigidity of the real-world objects can only be captured by going beyond 2-D shape models and resorting to fully 3-D models to describe the shape of the objects. To recover automatically the 3-D shape of the objects and the 3-D motion of the camera from the 2-D motion of the brightness pattern on the image plane, we describe in this chapter the surface-based rank 1 factorization method.

Content-based video representations, either single-layer or multiple-layer GV, or full 3-D object representations involve as an important preprocessing step the segmentation and tracking of 2-D objects. Segmentation is a very difficult problem for which there is a wealth of approaches described in the literature. We discuss in this chapter contour-based methods that are becoming popular. These methods are based on energy minimization approaches and extend beyond the well-known "snakes" method in which a set of points representing positions on the image boundary of 2-D objects — contours — is tracked in time. These methods make certain assumptions regarding the smoothness of these contours and how they evolve over time. These assumptions are at the heart of representing "active" contours. For completeness, we briefly discuss active-contour-based segmentation methods in this chapter.

In the next three subsections, we briefly overview work by others on single- and multilayered video representations and 3-D representations. Section 2.2 overviews active-contour-based approaches to segmentation. We then focus in Section 2.3 on generative video and its generalizations to multilayered representations and in Section 2.4 on the rank 1 surface-based 3-D video representations. Section 2.5 concludes the chapter.

2.1.1 Mosaics for static 3-D scenes and large depth: single layer

Image mosaics have received considerable attention from the fields of astronomy, biology, aerial photogrammetry, and image stabilization to video compression, visualization, and virtualized environments, among others. The main assumption in these application domains is that the 3-D scene layout is given by static regions shown very far away from the camera, that is, with large average depth values with respect to (w.r.t.) to the camera (center). Methods using this assumption will be discussed next.

Lippman [1] developed the idea of mosaics in the context of video production. This reference deals mostly with generating panoramic images describing static background regions. In this technique, panoramic images are generated by accumulating and integrating local image intensity information. Objects moving in the scene are averaged out; their shape and position in the image are described as a "halo" region containing the background region; the position of the object in the sequence is reconstructed by appropriately matching the background region in the halo to that of the background region in the enlarged image. Lippman's target application is high-definition television (HDTV) systems that require the presentation of video at different aspect ratios compared to standard TV. Burt and Adelson [2] describe a multiresolution technique for image mosaicing. Their aim is to generate photomosaics for which the region of spatial transition between different images (or image parts) is smooth in terms of its gray level or color difference. They use for this purpose Laplacian pyramid structures to decompose each image into their component pass-band images defined at different spatial resolution levels. For each band, they generate a mosaic, and the final mosaic is given by combining the mosaics at the different pass-bands. Their target applications are satellite imagery and computer graphics.

Hansen [3] and collaborators at the David Sarnoff Laboratory have developed techniques for generating mosaics in the framework of military reconnaissance, surveillance, and target detection. Their motivation is image stabilization for systems moving at high speeds and that use, among other things, video information. The successive images of these video sequences display little overlap, and they show, in general, a static 3-D scene and in some cases a single moving (target) object. Image or camera stabilization is extremely difficult under these circumstances. Hansen and coworkers use a mosaic-based stabilization technique by which a given image of the video sequence is registered to the mosaic built from preceding images of the sequence instead of just from the immediately preceding image. This mosaic is called the reference mosaic. It describes an extended view of a static 3-D terrain. The sequential mosaic generation is realized through a series of image alignment operations, which include the estimation of global image velocity and of image warping.

Teodosio and Bender [4] have proposed salient video stills as a novel way to represent videos. A salient still represents the video sequence by a single high-resolution image by translating, scaling, and warping images of the sequence into a single high-resolution raster. This is realized by (i) calculating the optical flow between successive images; (ii) using an affine representation of the optical flow to appropriately translate, scale, and warp images; and (iii) using a weighted median of the high-resolution image. As an intermediate step, a continuous space–time raster is generated in order to appropriately align all pixels, regardless of whether the camera pans or zooms, thus creating the salient still.

Irani et al. [5] propose a video sequence representation in terms of static, dynamic, and multiresolution mosaics. A static mosaic is built from collections of "submosaics," one for each scene subsequence, by aligning all of its frames to a fixed coordinate system. This type of mosaic can handle cases of static scenes, but it is not adequate for one having temporally varying information. In the latter case, a dynamic mosaic is built from a collection of evolving mosaics. Each of these temporarily updated mosaics is updated according to information from the most recent frame. One difference with static mosaic generation is that the coordinate system of the dynamic mosaics can be moving with the current frame. This allows for an efficient updating of the dynamic content.

2.1.2 Mosaics for static 3-D scenes and variable depth: multiple layers

When a camera moves in a static scene containing fixed regions or objects that cluster at different depth levels, it is necessary to generate multiple mosaics, one for each layer.

Wang and Adelson [6] describe a method to generate layers of panoramic images from video sequences generated through camera translation with respect to static scenes. They use the information from the induced (camera) motion. They segment the panoramic images into layers according to the motion induced by the camera motion. Video mosaicing is pixel based. It generates panoramic images from static scenery panned or zoomed by a moving camera.

2.1.3 Video representations with fully 3-D models

The mosaicing approaches outlined above represent a video sequence in terms of flat scenarios. Since the planar mosaics do not model the 3-D shape of the objects, these approaches do not provide a clear separation among object shape, motion, and texture. Although several researchers proposed enhancing the mosaics by incorporating depth information (see, for example, the plane + parallax approach [5, 7]), these models often do not provide meaningful representations for the 3-D shape of the objects. In fact, any video sequence obtained by rotating the camera around an object demands a content-based representation that must be fully 3-D based.

Among 3-D-model-based video representations, the semantic coding approach assumes that detailed *a priori* knowledge about the scene is available. An example of semantic coding is the utilization of head-and-shoulders parametric models to represent facial video sequences (see [8, 9]). The video analysis task estimates along time the small changes of the head-and-shoulders model parameters. The video sequence is represented by the sequence of estimated head-and-shoulders model parameters. This type of representation enables very high compression rates for the facial video sequences but cannot cope with more general videos.

The use of 3-D-based representations for videos of general scenes demands the automatic 3-D modeling of the environment. The information source for a number of successful approaches to 3-D modeling has been a range image (see, for example, [10, 11]).

This image, obtained from a range sensor, provides the depth between the sensor and the environment facing it on a discrete grid. Since the range image itself contains explicit information about the 3-D structure of the environment, the references cited above deal with the problem of how to combine a number of sets of 3-D points (each set corresponding to a range image) into a 3-D model.

When no explicit 3-D information is given, the problem of computing automatically a 3-D-model-based representation is that of building the 3-D models from the 2-D video data. The recovery of the 3-D structure (3-D shape and 3-D motion) of rigid objects from 2-D video sequences has been widely considered by the computer vision community. Methods that infer 3-D shape from a single frame are based on cues such as shading and defocus. These methods fail to give reliable 3-D shape estimates for unconstrained real-world scenes. If no prior knowledge about the scene is available, the cue to estimating the 3-D structure is the 2-D motion of the brightness pattern in the image plane. For this reason, the problem is generally referred to as structure from motion (SFM).

2.1.3.1   *Structure from motion: factorization*

Among the existing approaches to the multiframe SFM problem, the factorization method [12] is an elegant method to recover structure from motion without computing the absolute depth as an intermediate step. The object shape is represented by the 3-D position of a set of feature points. The 2-D projection of each feature point is tracked along the image sequence. The 3-D shape and motion are then estimated by factorizing a measurement matrix whose columns are the 2-D trajectories of each of the feature point projections. The factorization method proved to be effective when processing videos obtained in controlled environments with a relatively small number of feature points. However, to provide dense depth estimates and dense descriptions of the shape, this method usually requires hundreds of features, a situation that then poses a major challenge in tracking these features along the image sequence and that leads to a combinatorially complex correspondence problem.

In Section 2.4, we describe a 3-D-model-based video representation scheme that overcomes this problem by using the surface-based rank 1 factorization method [13, 14]. There are two distinguishing features of this approach. First, it is *surface* based rather than feature (point) based; i.e., it describes the shape of the object by patches, e.g., planar patches or higher-order polynomial patches. Planar patches provide not only localization but also information regarding the orientation of the surface. To obtain similar quality descriptions of the object, the number of patches needed is usually much smaller than the number of feature points needed. In [13], it is shown that the polynomial description of the patches leads to a parameterization of the object surface and this parametric description of the 3-D shape induces a parametric model for the 2-D motion of the brightness pattern in the image plane. Instead of tracking pointwise features, this method tracks regions of many pixels, where the 2-D image motion of each region is described by a single set of parameters. This approach avoids the correspondence problem and is particularly suited for practical scenarios in which the objects are, for example, large buildings that are well described by piecewise flat surfaces. The second characteristic of the method in [13, 14] and in Section 2.4 is that it requires only the factorization of a rank 1 rather than rank 3 matrix, which simplifies significantly the computational effort of the approach and is more robust to noise.

Clearly, the generation of images from 3-D models of the world is a subject that has been addressed by the computer graphics community. When the world models are inferred from photograph or video images, rather than specified by an operator, the view generation process is known as image-based rendering (IBR). Some systems use a set of calibrated cameras (i.e., with known 3-D positions and internal parameters) to capture the 3-D shape of the scene and synthesize arbitrary views by texture mapping, e.g., the Virtualized Reality system [15]. Other systems are tailored to the modeling of specific 3-D objects like the Façade system [16], which does not need *a priori* calibration but requires user interaction to establish point correspondences. These systems, as well as the framework described in Section 2.4, represent a scene by using geometric models of the 3-D objects. A distinct approach to IBR uses the plenoptic function [17] — an array that contains the light intensity as a function of the viewing point position in 3-D space, the direction of propagation, the time, and the wavelength. If in empty space, the dependence on the viewing point position along the direction of propagation may be dropped. By dropping also the dependence on time, which assumes that the lighting conditions are fixed, researchers have attempted to infer from images what has been called the light field [18]. A major problem in rendering images from acquired light fields is that, due to limitations on the number of images available and on the processing time, they are usually subsampled. The Lumigraph system [19] overcomes this limitation by using the approximate geometry of the 3-D scene to aid the interpolation of the light field.

2.2 Image segmentation

In this section, we discuss segmentation algorithms, in particular, energy minimization and active-contour-based approaches, which are popularly used in video image processing. In Subsection 2.2.1, we review concepts from variational calculus and present several forms of the Euler-Lagrange equation. In Subsection 2.2.2, we broadly classify the image segmentation algorithms into two categories: edge-based and region-based. In Subsection 2.2.3, we consider active contour methods for image segmentation and discuss their advantages and disadvantages. The seminal work on active contours by Kass, Witkin, and Terzopoulos [20], including its variations, is then discussed in Subsection 2.2.4. Next, we provide in Subsection 2.2.5 background on curve evolution, while Subsection 2.2.6 shows how curve evolution can be implemented using the level set method. Finally, we provide in Subsection 2.2.7 examples of segmentation by these geometric active contour methods utilizing curve evolution theory and implemented by the level set method.

2.2.1 Calculus of variations

In this subsection, we sketch the key concepts we need from the calculus of variations, which are essential in the energy minimization approach to image processing. We present the Euler-Lagrange equation, provide a generic solution when a constraint is added, and, finally, discuss gradient descent numerical solutions.

Given a scalar function $u(x):[0,1] \to \mathbf{R}$ with given constant boundary conditions $u(0)=a$ and $u(1)=b$, the basic problem in the calculus of variations is to minimize an energy functional [21]

$$J(u) = \int_0^1 E(u(x), u'(x)) dx, \qquad (2.1)$$

where $E(u,u')$ is a function of u and u', the first derivative of u. From classical calculus, we know that the extrema of a function $f(x)$ in the interior of the domain are attained at the zeros of the first derivative of $f(x)$, i.e., where $f'(x) = 0$. Similarly, to find the extrema of the functional $J(u)$, we solve for the zeros of the first variation of J, i.e., $\delta J = 0$. Let δu and $\delta u'$ be small perturbations of u and u', respectively. By Taylor series expansion of the integrand in Equation (2.1), we have

$$E(u + \delta u, u' + \delta u') = E(u, u') + \frac{\partial E}{\partial u} \delta u + \frac{\partial E}{\partial u'} \delta u' + \cdots. \qquad (2.2)$$

Then

$$J(u + \delta u) = J(u) + \int_0^1 \left(E_u \delta u + E_{u'} \delta u' \right) dx + \cdots, \qquad (2.3)$$

where $E_u = \dfrac{\partial E}{\partial u}$ and $E_{u'} = \dfrac{\partial E}{\partial u'}$ represent the partial derivatives of $E(u,u')$ with respect to u and u', respectively. The first variation of J is then

$$\delta J(u) = J(u + \delta u) - J(u) \qquad (2.4)$$

$$= \int_0^1 \left(E_u \delta u + E_{u'} \delta u' \right) dx. \qquad (2.5)$$

Integrating by parts the second term of the integral, we have

$$\int_0^1 E_{u'} \delta u' \, dx = E_{u'} \delta u(x) \big|_{x=0}^{x=1} - \int_0^1 \delta u \frac{d}{dx} \left(E_{u'} \right) dx \qquad (2.6)$$

$$= -\int_0^1 \delta u \frac{d}{dx} \left(E_{u'} \right) dx. \qquad (2.7)$$

The nonintegral term vanishes because $\delta u(1) = \delta u(0) = 0$ due to the assumed constant boundary conditions of u. Substituting Equation (2.7) back into Equation (2.4), we obtain

$$\delta J(u) = \int_0^1 \left[\delta u \, E_u - \delta u \frac{d}{dx}(E_{u'}) \right] dx. \tag{2.8}$$

A necessary condition for u to be an extremum of $J(u)$ is that u makes the integrand zero, i.e.,

$$E_u - \frac{d}{dx} E_{u'} = \frac{\partial E}{\partial u} - \frac{d}{dx}\left(\frac{\partial E}{\partial u'}\right) = 0. \tag{2.9}$$

This is the Euler-Lagrange equation for a one-dimensional (1-D) problem in the calculus of variations [21].

More generally, the Euler-Lagrange equation for an energy functional of the form

$$J(u) = \int_0^1 E(x, u, u', u'', \cdots, u^n) dx, \tag{2.10}$$

where u^n is the nth derivative of $u(x)$ with respect to x, can be derived in a similar manner as

$$E_u - \frac{d}{dx} E_{u'} + \frac{d^2}{dx^2} E_{u''} -, \cdots, +(-1)^n \frac{d^n}{dx^n} E_{u^n} = 0. \tag{2.11}$$

For a scalar function defined on a 2-D domain or a 2-D plane, $u(x, y): \mathbf{R}^2 \to \mathbf{R}$, we have a similar result. For instance, given an energy functional

$$J(u) = \iint_\Omega E(u, u_x, u_y, u_{xx}, u_{yy}) dx \, dy, \tag{2.12}$$

the corresponding Euler-Lagrange equation is given by

$$\frac{\partial E}{\partial u} - \frac{d}{dx}\left(\frac{\partial E}{\partial u_x}\right) - \frac{d}{dy}\left(\frac{\partial E}{\partial u_y}\right) + \frac{d^2}{dx^2}\left(\frac{\partial E}{\partial u_{xx}}\right) + \frac{d^2}{dy^2}\left(\frac{\partial E}{\partial u_{yy}}\right) = 0. \tag{2.13}$$

Analogously, we obtain a system of Euler-Lagrange equations for a vector-value function $\mathbf{u}$. For example, if $\mathbf{u}(x) = [u_1(x) \ u_2(x)]^T : \mathbf{R} \to \mathbf{R}^2$, then the corresponding system of Euler-Lagrange equations is

$$E_{u_1} - \frac{d}{dx}E_{u_1'} + \frac{d^2}{dx^2}E_{u_1''} - , \cdots , + (-1)^n \frac{d^n}{dx^n}E_{u_1^n} = 0, \tag{2.14}$$

$$E_{u_2} - \frac{d}{dx}E_{u_2'} + \frac{d^2}{dx^2}E_{u_2''} - , \cdots , + (-1)^n \frac{d^n}{dx^n}E_{u_2^n} = 0. \tag{2.15}$$

2.2.1.1 Adding constraints

Usually, we are not allowed to freely search for the optimal u; rather, constraints are added. For instance, we may want to search for a function $u(x)$ that minimizes the energy functional

$$J_1(u) = \int_a^b E(x,u,u')dx, \tag{2.16}$$

under a constraint functional

$$J_2(u) = \int_a^b G(x,u,u')dx = c, \tag{2.17}$$

where c is a given constant. By use of a Lagrange multiplier λ, the new energy functional becomes

$$J(u) = J_1(u) - \lambda J_2(u) \tag{2.18}$$

$$= \int_a^b \left[E(x,u,u') - \lambda G(x,u,u') \right] dx. \tag{2.19}$$

As a result, the corresponding Euler-Lagrange equation is

$$\frac{\partial E}{\partial u} - \frac{d}{dx}E_{u'} - \lambda \left(\frac{\partial G}{\partial u} - \frac{d}{dx}G_{u'} \right) = 0, \tag{2.20}$$

which must be solved subject to the constraint Equation (2.17).

2.2.1.2 Gradient descent flow

One of the fundamental questions in the calculus of variations is how to solve the Euler-Lagrange equation, i.e., how to solve for u in

$$\mathcal{F}(u) = 0, \tag{2.21}$$

where $\mathcal{F}(u)$ is a generic function of u whose zero makes the first variation of a functional J zero, i.e., $\delta J = 0$. Equation (2.21) can be any of the Euler-Lagrange equations in (2.11), (2.13), (2.14), or (2.20). Only in a very limited number of simple cases is this problem solved analytically. In most image-processing applications, directly solving this problem is infeasible. One possible solution for $\mathcal{F}(u) = 0$ is to first let $u(x)$ be a function of an(other) artificial time marching parameter t and then numerically solve the partial differential equation (PDE)

$$\frac{\partial u}{\partial t} = \mathcal{F}(u), \tag{2.22}$$

with a given initial $u_0(x)$ at $t = 0$. At steady state,

$$\frac{\partial u}{\partial t} = 0 \tag{2.23}$$

implies that $\mathcal{F}(u) = 0$ is achieved, and the solution to the Euler-Lagrange equation is obtained. This is denoted as the gradient descent flow method.

2.2.2 Overview of image segmentation methods

Image segmentation is a fundamental step in building extended images, as well as many other image- and video-processing techniques. The principal goal of image segmentation is to partition an image into clusters or regions that are homogeneous with respect to one or more characteristics or features. The first major challenge in segmenting or partitioning an image is the determination of the defining features that are unique to each meaningful region so that they may be used to set that particular region apart from the others. The defining features of each region manifest themselves in a variety of ways, including, but not limited to, image intensity, color, surface luminance, and texture. In generative video and structure from motion, an important feature is the 2-D-induced motion of the feature points or the surface patches. Once the defining features are determined, the next challenging problem is how to find the "best" way to capture these defining features through some means such as statistical characteristics, transforms, decompositions, or other more complicated methodologies, and then use them to partition the image efficiently. Furthermore, any corruption — by noise, motion artifacts, and the missing data due to occlusion within the observed image — poses additional problems to the segmentation process. Due to these difficulties, the image segmentation problem remains a significant and considerable challenge.

The image segmentation algorithms proposed thus far in the literature may be broadly categorized into two different approaches, each with its own strengths and weaknesses [22, 23]:

2.2.2.1   Edge-based approach

The edge-based approach relies on discontinuity in image features between distinct regions. The goal of edge-based segmentation algorithms is to locate the object boundaries, which separate distinct regions, at the points where the image has high change (or gradient) in feature values. Most edge-based algorithms exploit spatial information by examining local edges found within the image. They are often very easy to implement and quick to compute, as they involve a local convolution of the observed image with a gradient filter. Moreover, they do not require *a priori* information about image content. The Sobel [24], Prewitt [25], Laplacian [26, 27], or Canny [28] edge detectors are just a few examples. For simple noise-free images, detection of edges results in straightforward boundary delineation. However, when applied to noisy or complex images, edge detectors have three major problems:

1. They are very sensitive to noise.
2. They require a selection of an edge threshold.
3. They do not generate a complete boundary of the object because the edges often do not enclose the object completely due to noise or artifacts in the image or the touching or overlapping of objects.

These obstacles are difficult to overcome because solving one usually leads to added problems in the others. To reduce the effect of the noise, one may lowpass filter the image before applying an edge operator. However, lowpass filtering also suppresses soft edges, which in turn leads to more incomplete edges to distinguish the object boundary. On the other hand, to obtain more complete edges, one may lower the threshold to be more sensitive to, and thus include more, weak edges, but this means more spurious edges appear due to noise. To obtain satisfactory segmentation results from edge-based techniques, an ad hoc postprocessing method such as the vector graph method of Casadei and Mitter [29, 30] is often required after the edge detection to link or group edges that correspond to the same object boundary and get rid of other spurious edges. However, such an automatic edge linking algorithm is computationally expensive and generally not very reliable.

2.2.2.2   Region-based approach

The region-based approach, as opposed to the edge-based approach, relies on the similarity of patterns in image features within a cluster of neighboring pixels. Region-based techniques, such as region growing or region merging [31, 32, 33], assign membership to objects based on homogeneous statistics. The statistics are generated and updated dynamically. Region-growing methods generate a segmentation map by starting with small regions that belong to the structure of interest, called seeds. To grow the seeds into larger regions, the neighboring pixels are then examined one at a time. If they are sufficiently similar to the seeds, based on a uniformity test, then they are assigned to the growing region. The procedure continues until no more pixels can be added. The seeding scheme to create the initial regions and the homogeneity criteria for when and how to merge regions are determined *a priori*. The advantage of region-based models is that the statistics of the entire image, rather than local image information, are considered. As a result, the techniques are robust to noise and can be used to locate boundaries that do not correspond to large image gradients. However, there is no provision in the region-based framework to include the object boundary in the decision-making process, which usually leads to irregular or noisy boundaries and holes in the interior of the object. Moreover, the seeds have to be initially picked (usually by an operator) to be within the region of interest, or else the result may be undesirable.

2.2.3   Active contour methods

Among a wide variety of segmentation algorithms, active contour methods [20, 34–42] have received considerable interest, particularly in the video image–processing community. The first active contour method, called "snake," was introduced in 1987 by Kass, Witkin, and Terzopoulos [20, 34]. Since then the techniques of active contours for image segmentation have grown significantly and have been used in other applications as well. An extensive discussion of various segmentation methods as well as a large set of references on the subject may be found in [43].

Because active contour methods deform a closed contour, this segmentation technique guarantees continuous closed boundaries in the resulting segmentation. In principle, active contour methods involve the evolution of curves toward the boundary of an object through the solution of an energy functional minimization problem. The energy functionals in active contour models depend not only on the image properties but also on the shape of the contour. Therefore, they are considered a high-level image segmentation scheme, as opposed to the traditional low-level schemes such as edge detectors [24, 28] or region-growing methods [31, 32, 33].

The evolution of the active contours is often described by a PDE, which can be tracked either by a straightforward numerical scheme such as the Lagrangian parameterized control points [44] or by more sophisticated numerical schemes such as the Eulerian level set methods [45, 46].

Although traditional active contours for image segmentation are edge based, the current trends are region-based active contours [40, 42] or hybrid active contour models, which utilize both region-based and edge-based information [39, 41]. This is because the region-based models, which rely on regional statistics for segmentation, are more robust to noise and less sensitive to the placement of the initial contour than the edge-based models.

The classical snake algorithm [20] works explicitly with a parameterized curve. Thus, it is also referred to as a parametric active contour, in contrast to the geometric active contour [47], which is based on the theory of curve evolution. Unlike the parametric active contour methods, the geometric active contour methods are usually implemented implicitly through level sets [45, 46].

In the following subsections, we describe the parametric active contour method, or classical snakes, and discuss its advantages and its shortcomings in Subsection 2.2.4. We also present two variations of classical snakes that attempt to improve the snake algorithms. We then provide background on the contour evolution theory and the level set method in Subsections 2.2.5 and 2.2.6, respectively. We finally show in Subsection 2.2.7 how the geometric contour method, which is based on the curve evolution theory and often implemented by the level set method, can improve the performance of image segmentation over the parametric active contour-based algorithms.

2.2.4   *Parametric active contour*

The parametric active contour model or snake algorithm [20] was first introduced in the computer vision community to overcome the traditional reliance on low-level image features like pixel intensities. The active contour model is considered a high-level mechanism because it imposes the shape model of the object in the processing. The snake algorithm turns the boundary extraction problem into an energy minimization problem [48]. A traditional snake is a parameterized curve $\mathbf{C}(p) = [\, x(p) \;\; y(p)\,]^{T}$ for $p \in [0,1]$ that moves through a spatial domain Ω of the image $I(x,y)$ to minimize the energy functional

$$J(\mathbf{C}) = \varepsilon_{int}(\mathbf{C}) + \varepsilon_{ext}(\mathbf{C}). \tag{2.24}$$

It has two energy components, the internal energy ε_{int} and the external energy ε_{ext}. The high-level shape model of the object is controlled by the internal energy, whereas the external energy is designed to capture the low-level features of interest, very often edges. The main idea is to minimize these two energies simultaneously. To control the smoothness and the continuity of the curve, the internal energy governs the first and second derivatives of the contour, i.e.,

$$\varepsilon_{int} = \frac{1}{2} \int_{0}^{1} \alpha \big| \mathbf{C}'(p) \big|^{2} + \beta \big| \mathbf{C}''(p) \big|^{2} \, dp, \tag{2.25}$$

where α and β are constants and $\mathbf{C}'(p)$ and $\mathbf{C}''(p)$ are the first and second derivatives of the contour with respect to the indexing variable p, respectively. The first derivative discourages stretching and makes the contour behave like an elastic string. The second derivative discourages bending and makes it behave like a rigid rod. Therefore, the weighting parameters α and β are used to control the strength of the model's elasticity and rigidity, respectively.

The external energy, on the other hand, is computed by integrating a potential energy function $P(x,y)$ along the contour $C(p)$, i.e.,

$$\mathcal{E}_{ext} = \int_0^1 P(C(p))\, dp, \tag{2.26}$$

where $P(x,y)$ is derived from the image data. The potential energy function $P(x,y)$ must take small values at the salient features of interest because the contour $C(p)$ is to search for the minimum external energy. Given a gray-level image $I(x,y)$, viewed as a function of the continuous variables (x,y), a typical potential energy function designed for the active contour C that captures step edges is

$$P(x,y) = -|\nabla G_\sigma(x,y) * I(x,y)|^2, \tag{2.27}$$

where $G_\sigma(x,y)$ is a 2-D Gaussian function with variance σ^2, ∇ represents the gradient operator, and $*$ is the image convolution operator. The potential energy function defined as in (2.27) is called the *edge map* of the image. Figure 2.1(b) shows the corresponding edge map of the image in Figure 2.1(a).

The problem of finding a curve $C(p)$ that minimizes an energy functional $J(C(p))$ is known as a variational problem [21]. It has been shown in [20] that the curve C that minimizes $J(C)$ in (2.24) must satisfy the following Euler-Lagrange equation

$$\alpha C''(p) - \beta C''''(p) - \nabla P(C(p)) = 0. \tag{2.28}$$

To find a solution to Equation (2.28), the snake is made dynamic by first letting the contour $C(p)$ be a function of time t (as well as p), i.c., $C(p,t)$, and then replacing the **0** on the right-hand side of Equation (2.28) by the partial derivative of C with respect to t as the following

$$\frac{\partial C}{\partial t} = \alpha C''(p) - \beta C''''(p) - \nabla P(C(p)). \tag{2.29}$$

The gradient descent method is then used to iteratively solve for the zero of (2.29).

To gain some insight about the physical behavior of the evolution of active contours, Xu and Prince realized Equation (2.29) as the balancing between two forces [38]

$$\frac{\partial C}{\partial t} = F_{int}(C) + F_{ext}(C), \tag{2.30}$$

where the internal force is given by

$$\mathbf{F}_{int} = \alpha\mathbf{C}''(p) - \beta\mathbf{C}''''(p), \tag{2.31}$$

and the external force is given by

$$\mathbf{F}_{ext} = -\nabla P(x, y). \tag{2.32}$$

The internal force $\mathbf{F}_{int}$ dictates the regularity of the contour, whereas the external force $\mathbf{F}_{ext}$ pulls it toward the desired image feature. We call $\mathbf{F}_{ext}$ the *potential force field*, because it is the vector field that pulls the evolving contour toward the desired feature (edges) in the image. Figure 2.1(c) shows the potential force field, which is the negative gradient magnitude of the edge map in Figure 2.1(b). Figure 2.1(d) zooms in the area within the square box shown in Figure 2.1(c).

The snake algorithm gains its popularity in the computer vision community because of the following characteristics:

1. It is deformable, which means it can be applied to segment objects with various shapes and sizes.
2. It guarantees a smooth and closed boundary of the object.
3. It has been proven very useful in motion tracking for video.

The major drawbacks associated with the snake's edge-based approach are:

1. It is very sensitive to noise because it requires the use of differential operators to calculate the edge map.
2. The potential forces in the potential force field are only present in the close vicinity of high values in the edge map.
3. It utilizes only the local information along the object boundaries, not the entire image.

Hence, for the snake algorithm to converge to a desirable result, the initial contour must be placed close enough to the true boundary of the object. Otherwise, the evolving contour might stop at undesirable spurious edges or the contour might not move at all if the potential force on the contour front is not present. As a result, the initial contour is often obtained manually. This is a key pitfall of the snake method.

2.2.4.1 Variations of classical snakes

Many efforts have been made to address the limitations of the original snakes method. For example, to help the snake move or avoid being trapped by spurious isolated edge points, Cohen's balloon snake approach [35] added another artificial inflation force to the external force component of Equation (2.30). Thus, the balloon snake's external force becomes

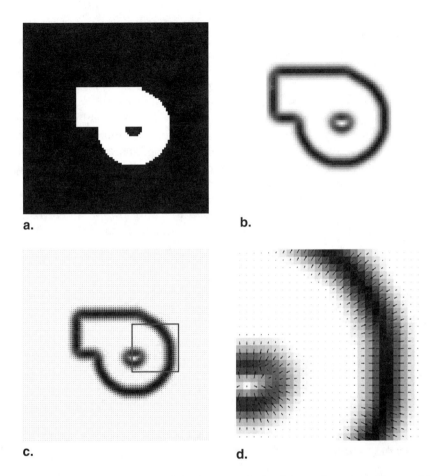

Figure 2.1 (a) Original image; (b) edge map derived from the original image (a), (c) potential force field: the negative gradient of the edge map (b); (d) zoom-in of area within the square box in (c).

$$\mathbf{F}_{ext} = -\nabla P(x,y) + F_{const}\, \hat{\mathbf{n}}, \tag{2.33}$$

where F_{const} is an arbitrary constant and $\hat{\mathbf{n}}$ is the unit normal vector on the contour front. However, the balloon snake has limitations. Although the balloon snake aims to pass through edges that are too weak with respect to the inflation force $F_{const}\, \hat{\mathbf{n}}$, adjusting the strength of the balloon force is difficult because it must be large enough to overcome the weak edges and noises but small enough not to overwhelm a legitimate boundary. Besides, the balloon force is image independent; i.e., it is not derived from the image. Therefore, the contour will continue to inflate at the points where the true boundary is missing or weaker than the inflation force.

Xu and Prince [38, 49] introduced a new external force for edge-based snakes called the gradient vector flow (GVF) snake. In their method, instead of directly using the gradient of the edge map as the potential force field, they diffuse it first to obtain a new force field that has a larger capture range than the gradient of the edge map. Figures 2.2(a) and (b) depict the gradient of an edge map and the Xu and Prince's new force field, respectively. Comparing the two figures, we observe that the Xu and Prince's vector forces gradually decrease as they are away from the edge pixels, whereas the vector forces in the gradient of the edge map exist only in the neighboring pixels of the edge pixels. As a result, there are no forces to pull a contour that is located at the pixels far away from the edge pixels in the gradient of the edge map field, but the contour may experience some forces at the same location in the Xu and Prince's force field.

Two other limitations associated with the parametric representation of the classical snake algorithm are the need to perform reparameterization and topological adaptation. It is often necessary to dynamically reparameterize the snake in order to maintain a faithful delineation of the object boundary. This adds computational overhead to the algorithm. In addition, when the contour fragments need to be merged or split, it may require a new topology and, thus, the reconstruction of the new parameterization. McInerney and Terzopoulos [50] have proposed an algorithm to address this problem.

2.2.5   Curve evolution theory

In this subsection, we explain how to control the motion of a propagating contour using the theory of curve evolution. In particular, we present two examples of the motion for a propagating curve that are commonly used in active contour schemes for image segmentation.

Denote a family of smooth contours as

$$\mathbf{C}(p,t) = \begin{bmatrix} x(p,t) \\ y(p,t) \end{bmatrix}, \qquad (2.34)$$

where $p \in [0,1]$ parameterizes the set of points on each curve, and $t \in [0,\infty)$ parameterizes the family of curves at different time evolutions. With this parameterization scheme, a closed contour has the property that

$$\mathbf{C}(0,t) = \mathbf{C}(1,t) \quad \forall t. \qquad (2.35)$$

We are interested in finding an equation for the propagating motion of a curve that eventually segments the image. Assume a variational approach for image segmentation formulated as finding the curve $\mathbf{C}^*$ such that

$$\mathbf{C}^* = \underset{\mathbf{C}}{\operatorname{argmin}} \ J(\mathbf{C}), \qquad (2.36)$$

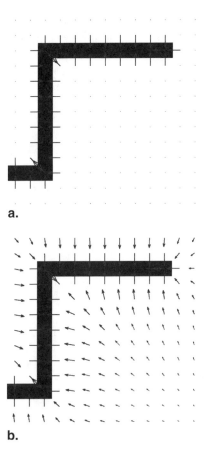

a.

b.

Figure 2.2 Two examples of the potential force fields of an edge map: (a) gradient of the edge map; (b) Xu and Prince's GVF field.

where J is an energy functional constructed to capture all the criteria that lead to the desired segmentation. The solution to this variational problem often involves a PDE.

Let $F(C)$ denote an Euler-Lagrange equation such that the first variation of $J(C)$ with respect to the contour C is zero. Under general assumptions, the necessary condition for C to be the minimizer of $J(C)$ is that $F(C) = 0$. The solution to this necessary condition can be computed as the steady state solution of the following PDE [51]

$$\frac{\partial C}{\partial t} = F(C). \tag{2.37}$$

This equation is the curve evolution equation or the *flow* for the curve C. The form of this equation indicates that $F(C)$ represents the "force" acting upon the contour front. It can also be viewed as the velocity at which the contour evolves. Generally, the force **F** has two components. As depicted in Figure 2.3, **F**$_{\hat{n}}$ is the component of **F** that points in the normal direction with respect to the contour front, and **F**$_t$ is the (other) component of **F** that is tangent to **C**.

In curve evolution theory, we are interested only in **F**$_{\hat{n}}$ because it is the force that moves the contour front forward (or inward), hence changing the geometry of the contour. The flow along **F**$_t$, on the other hand, only reparameterizes the curve and does not play any role in the evolution of the curve. Therefore, the curve evolution equation is often reduced to just the normal component as

$$\frac{\partial \mathbf{C}}{\partial t} = F\,\hat{\mathbf{n}}, \tag{2.38}$$

where F is called the speed function. In principle, the speed function depends on the local and global properties of the contour. Local properties of the contour include local geometric information such as the contour's principal curvature κ or the unit normal vector $\hat{n}$ of the contour. Global properties of the curve depend on its shape and position.

Coming up with an appropriate speed function, or equivalently the curve evolution equation, for the image segmentation underlies much of the research in this field. As an example, consider the Euclidean curve-shortening flow given by

$$\frac{\partial \mathbf{C}}{\partial t} = \kappa\,\hat{\mathbf{n}}. \tag{2.39}$$

This flow corresponds to the gradient descent along the direction in which the Euclidean arc length of the curve

Figure 2.3 The normal and tangential components of a force on the contour front.

$$L = \oint_C ds \qquad (2.40)$$

decreases most rapidly. As shown in Figure 2.4, a jagged closed contour evolving under this flow becomes smoother. Flow (2.39) has a number of attractive properties, which make it very useful in a range of image-processing applications. However, it is never used alone because if we continue the evolution with this flow, the curve will shrink to a circle, then to a point, and then finally vanishes.

Another example illustrates some of the problems associated with a propagating curve. Consider the curve-evolution equation

$$\frac{\partial \mathbf{C}}{\partial t} = V_o \hat{\mathbf{n}}, \qquad (2.41)$$

where V_o is a constant. If V_o is positive, the contour inflates. If V_o is negative, the contour evolves in a deflationary fashion. This is because it corresponds to the minimization of the area within the closed contour.

As seen in Figure 2.5, most curves evolving under the constant flow (2.41) often develop sharp points or corners that are nondifferentiable (along the contour). These singularities pose a problem of how to continue implementing the next evolution of the curve because the normal to the curve at a singular point is ambiguous. However, an elegant numerical implementation through the level set method provides an "entropy solution" that solves this curve evolution problem [45, 46, 52, 53]. Malladi et al. [37] and Caselles et al. [54] utilized both the curvature flow (2.39) and the constant flow (2.41) in their active contour schemes for image segmentation because they are complementary to each other. Whereas the constant flow can create singularities from an initial smooth contour, the curvature flow removes them by smoothing the contour in the process.

Figure 2.4 Flow under curvature: a jagged contour becomes smoother.

Figure 2.5 Flow with negative constant speed deflates the contour.

2.2.6   Level set method

Given a current position for the contour **C** and the equation for its motion such as the one in (2.37), we need a method to track this curve as it evolves. In general, there are two approaches to track the contour, the Lagrangian and the Eulerian approaches. The Lagrangian approach is a straightforward difference approximation scheme. It parameterizes the contour discretely into a set of control points lying along the moving front. The motion vectors, derived from the curve-evolution equation through a difference approximation scheme, are then applied to these control points to move the contour front. The control points then advance to their new locations to represent the updated curve front. Though this is a natural approach to track the evolving contour, the approach suffers from several problems [55]:

1. This approach requires an impractically small time step to achieve a stable evolution.
2. As the curve evolves, the control points tend to "clump" together near high curvature regions, causing numerical instability. Methods for control points reparameterization are then needed, but they are often less than perfect and hence can give rise to errors.
3. Besides numerical instability, there are also problems associated with the way the Lagrangian approach handles topological changes. As the curve splits or merges, topological problems occur, requiring ad hoc techniques [50, 56] to continue to make this approach work.

Osher and Sethian [45, 46, 52, 53] developed the level set technique for tracking curves in the Eulerian framework, written in terms of a fixed coordinate system. There are four main advantages to this level set technique:

1. Since the underlying coordinate system is fixed, discrete mesh points do not move; the instability problems of the Lagrangian approximations can be avoided.
2. Topological changes are handled naturally and automatically.
3. The moving front is accurately captured regardless of whether it contains cusps or sharp corners.

4. The technique can be extended to work on any number of spatial dimensions.

The level set method [45] implicitly represents the evolving contour $\mathbf{C}(t)$ by embedding it as the zero level of a level set function $\phi : \mathbf{R}^2 \times [0, \infty) \to \mathbf{R}$, i.e.,

$$\mathbf{C}(t) = \left\{ (x,y) \in \Omega : \phi(x,y,t) = 0 \right\}. \tag{2.42}$$

Starting with an initial level set function $\phi(t=0)$, we then evolve $\phi(t)$ so that its zero level set moves according to the desired flow of the contour. Based on the convention that this level set graph has negative values inside $\mathbf{C}$ and positive values outside $\mathbf{C}$, i.e.,

$$\text{inside}(\mathbf{C}) = \Omega_1 = \left\{ (x,y) \in \Omega : \phi(x,y,t) > 0 \right\}, \tag{2.43}$$

$$\text{outside}(\mathbf{C}) = \Omega_2 = \left\{ (x,y) \in \Omega : \phi(x,y,t) < 0 \right\}, \tag{2.44}$$

the level set function ϕ can be implemented as the signed Euclidean distance to the contour $\mathbf{C}$. For details about how to implement the Euclidean distance to a contour, see [46, 57]. Using the standard Heaviside function

$$\mathcal{H}(\phi) = \begin{cases} 1, & \text{if } \phi \geq 0 \\ 0, & \text{if } \phi < 0' \end{cases} \tag{2.45}$$

we can conveniently mask out the image pixels that are inside, outside, or on the contour $\mathbf{C}$. For instance, the function $\mathcal{H}(\phi)$ represents the binary template of the image pixels that are inside or on the contour. The function $1 - \mathcal{H}(\phi)$ represents the binary template of the image pixels that are strictly outside the contour. To select only the pixels that are on the contour $\mathbf{C}$, we can use $\mathcal{H}(\phi) - [1 - \mathcal{H}(-\phi)]$. To facilitate numerical implementation, however, the regularized Heaviside function and its derivative, the regularized delta function, are often used instead. Define the regularized Heaviside function by

$$\mathcal{H}_\epsilon(\phi) = \frac{1}{2} \left[1 + \frac{2}{\pi} \arctan\left(\frac{\phi}{\epsilon} \right) \right] \tag{2.46}$$

then the regularized delta function is

$$\delta_\epsilon(\phi) = \frac{d}{d\phi} \mathcal{H}_\epsilon(\phi), \tag{2.47}$$

$$= \frac{1}{\pi} \left[\frac{\epsilon}{\epsilon^2 + \phi^2} \right].$$

(2.48)

The functions $\mathcal{H}_\epsilon(\phi)$, $1 - \mathcal{H}_\epsilon(\phi)$, and $\delta_\epsilon(\phi)$ are to represent the templates of the image pixels that are inside, outside, and on the contour $\mathbf{C}$, respectively. By defining the sign of the level set function ϕ to be positive inside and negative outside the contour, the unit normal vector $\mathbf{n}$ of the contour $\mathbf{C}$, defined as

$$\mathbf{n} = \frac{\nabla \phi}{|\nabla \phi|},$$

(2.49)

will point inward as shown in Figure 2.6. Furthermore, the curvature κ along the contour, defined as

$$\kappa = \mathrm{div}\,(\mathbf{n}) = \mathrm{div}\left(\frac{\nabla \phi}{|\nabla \phi|} \right) = \frac{\phi_{xx}\phi_y^2 - 2\phi_x\phi_y\phi_{xy} + \phi_{yy}\phi_x^2}{\left(\phi_x^2 + \phi_y^2 \right)^{3/2}}$$

(2.50)

is positive where the unit normal vectors diverge. On the other hand, the curvature of the contour is negative if the unit normal vectors converge (see Figure 2.6).

2.2.7   *Geometric active contours*

Based on the theory of curve evolution [58], geometric active contours [37, 47] evolve the curves using only geometric measures, such as the curvature and the normal vectors, resulting in a contour evolution that is independent of the curve's parameterization. Therefore, there is no need for reparameterization. In addition, the geometric active contour method can be implicitly

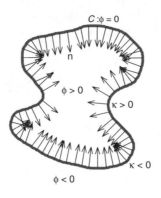

Figure 2.6 Unit normal vectors and curvature of the contour $\mathbf{C}$.

implemented by the level set technique [46], which handles the topology change automatically.

As mentioned before in Subsection 2.2.5, Malladi, Sethian, and Vemuri [37] and Caselles, Kimmel, and Sapiro [54] utilized curvature flow (2.39) and the constant flow (2.41) concurrently to move the contour **C** in the direction of its normal vector **n** as in

$$\frac{\partial \mathbf{C}}{\partial t} = -g \cdot (\kappa + V_o)\,\mathbf{n}, \qquad (2.51)$$

where κ is the curvature of the contour; V_o is a constant; and g, designed to capture prominent edges, is a decreasing function of the gradient image (or the edge map). An example of a suitable g is

$$g = \frac{1}{1+\left|\nabla G * I(x,y)\right|}. \qquad (2.52)$$

We can see that the contour front will slow down where the value of the edge map $\left|\nabla G * I(x,y)\right|$ is high because g approaches zero, but it will keep moving at constant speed V_o where the edge map value is zero. Therefore, the effect of the constant term V_o is the same as Cohen's balloon force [35]. As mentioned before, the effect of the curvature term κ is only to smooth out the contour [45]. Hence, it plays the role of the internal energy term in the classical snake method [20].

This scheme works well for objects that have well-defined edge maps. However, when the object boundary is difficult to distinguish, the evolving contour may leak out because the multiplicative term g only slows down the contour near the edges. It does not completely stop it. Chan and Vese [42] describe a new active contour scheme that does not use edges but combines an energy minimization approach with a level set-based solution. If **C** is a contour that partitions the domain of an image $I(x,y)$ into two regions, the region inside the contour Ω_1 and the region outside the contour Ω_2, then their approach is to minimize the functional

$$J(\mathbf{C}) = \mu \cdot Length(\mathbf{C})$$
$$+ \lambda_1 \int_{\Omega_1} \left|I(x,y) - c_1\right|^2 dxdy + \lambda_2 \int_{\Omega_2} \left|I(x,y) - c_2\right|^2 dxdy, \qquad (2.53)$$

where μ, λ_1, and λ_2 are constants and c_1 and c_2 are the average intensity values of the image pixels inside and outside the contour, respectively. As defined in (2.43) and (2.44), Ω_1 and Ω_2 represent the pixels inside and outside the contour **C**, respectively. If **C** is embedded as a zero level of the level set function ϕ, minimizing the functional (2.53) is equivalent to solving the PDE [42]

$$\frac{\partial \phi}{\partial t} = \left[\mu \operatorname{div}\left(\frac{\nabla \phi}{|\nabla \phi|} \right) - \lambda_1 (I - c_1)^2 + \lambda_2 (I - c_2)^2 \right] \delta_\epsilon(\phi), \qquad (2.54)$$

where div(·) denotes the divergence and $\delta_\epsilon(\phi)$ is the regularized delta function defined in (2.47). It masks out only the zero level set of ϕ, i.e., the contour **C**. The first term, the divergence term, which only affects the smoothness of the contour, is actually the motion under the curvature [45] because

$$\operatorname{div}\left(\frac{\nabla \phi}{|\nabla \phi|} \right) \delta(\phi) = \kappa(\mathbf{C}), \qquad (2.55)$$

where $\kappa(\mathbf{C})$ denotes the curvature of the contour **C**; see Equation (2.50). Therefore, Equation (2.54) is equivalent to the curve evolution

$$\frac{\partial \mathbf{C}}{\partial t} = \left[\mu \kappa - \lambda_1 (I - c_1)^2 + \lambda_2 (I - c_2)^2 \right] \mathbf{n}, \qquad (2.56)$$

where **n**, as defined in (2.49), is the outward unit normal vector of the contour **C**. The last two terms of (2.54), however, work together to move the contour such that all the pixels whose intensity values are similar to c_1 are grouped within the contour, and all the pixels whose intensity values are close to c_2 are assigned to be outside the contour.

As a result, the image will be segmented into two constant intensity regions. More importantly, unlike an edge-based model, this region-based geometric active contour model is less sensitive to the initial location of the contour. In addition, with the level set implementation, the algorithm handles the topological changes of the contour automatically when contour splitting or merging occurs. Figure 2.7(a) shows the process of segmenting a blurred-edge, piecewise-constant image with a hole in it using Chan and Vese's method, and Figure 2.7(b) depicts the final result. The initial contour, the dashed line, is simply an ellipse in the middle of the image and the final contours are shown as the two solid lines in Figure 2.7(b); one is the outside boundary of the object and the other is at the boundary of the hole within the object. We can see that this successful result is achieved even when the initial contour is far from the object's true boundary. Moreover, the splitting and merging of the contour around the hole, as seen in Figure 2.7(a), is done automatically through the level set implementation.

2.2.8 STACS: Stochastic active contour scheme

The segmentation approaches discussed in the previous subsections often work well but also fail in many important applications. For example, the approach in [42] can segment reasonably well an object from the background

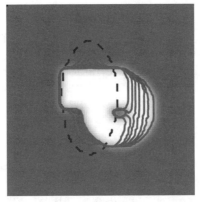

a.

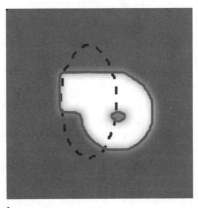

b.

Figure 2.7 Image segmentation using Chan and Vese's method: (a) evolving contour; (b) final result.

when the pixels inside and outside the contour follow the two-value model assumption well. In practice this is usually not the case, and this method may lead to poor segmentation results. Another common insufficiency of the segmentation methods presented in the previous subsections is that often we have a good indication of the shape of the object but the segmentation method has no explicit way to account for this knowledge. Edge- and region-based information may both be important clues, but existing methods usually take one or the other into account, not both.

We have developed in [59, 60] a method that we call the Stochastic Active Contour Scheme (STACS), which addresses these insufficiencies of existing approaches. It is an energy-based minimization approach in which the energy functional combines four terms: (i) an edge-based term; (ii) a region-based term that models the image textures stochastically rather than deterministically; (iii) a contour smoothness-based term; and, finally, (iv) a shape prior term. These four terms in the energy functional lead to very good segmentation results, even when the objects to be segmented exhibit low contrast with respect to neighboring pixels or the texture of different objects is quite similar. To enhance performance, STACS implements an annealing schedule on the relative weights among the four energy terms. This addresses the following issue. In practice, it is not clear which clue should dominate the segmentation; it is commonly true that, at the beginning of the segmentation process, edge- or region-based information is the more important clue, but, as the contour evolves, shape may become the dominant clue. STACS places initially more weight in the edge- and region-based terms but then slowly adapts the weights to reverse this relative importance so that, toward the end of the segmentation, more weight is placed in the shape prior and contour smoothness constraint terms. Lack of space prevents us from illustrating the good performance of STACS; details are in [59, 60] and references therein.

2.3   Mosaics: From 2-D to 3-D

In this section we will describe two approaches to mosaic generation. The first is called generative video [61, 62, 63, 64] and the second one [65, 66, 67] uses partial or full 3-D information to generate mosaics. The motivation in these two approaches is to overcome the limitations of classical mosaic generation methods; the objects in the (3-D) scene are very far from the camera so that there barely exists any parallax[2]. Depending on the relative geometry between the camera and the scene, different methods can be used to generate mosaics. For example, for objects far away from the camera, traditional mosaic generation methods are used. For incoming images of a video sequence $I_1, \cdots, I_N$, a mosaic M_k defined at time k is incrementally composed by combining the mosaic M_{k-1} with the current image I_k. Thus, in parallel to the video sequence, we can define a (partial) mosaic sequence $M_1, \cdots, M_N$. The spatial dimensions of this mosaic sequence change for each mosaic with time. The art of this composition has been explored in the last 25 years by using different blending techniques, i.e., the composition of M_k given M_{k-1} and I_k. Traditional application areas were astronomical, biological, and surveillance data. The need of more refined methods, which are used in robotics-related applications or immersive multimedia environments, required the processing of detailed 2-D and/or 3-D visual information, e.g., the segmentation of 2-D and/or 3-D objects. In this case, just knowledge of pure 2-D image information is not enough. We describe next how mosaics are generated in this case.

---

[2] Parallax describes the relative motion in the image plane of the projection of objects in (3-D) scenes: objects closer to the camera move at higher speed than objects further away.

2.3.1 Generative video

In generative video (GV) [61, 62, 63, 64], an input video sequence $I_1, \cdots, I_N$ is transformed into a set of constructs $C_1, \cdots, C_M$. These constructs (i) generalize the concept of mosaics to that of background (static) part of the image (scene) and foreground objects; (ii) use object stratification information, i.e., how they are organized in layers according to their relative 3-D depth information; and (iii) encode (2-D) object shape and velocity information. The goal is to achieve a compact representation of content-based video sequence information. The mosaics in GV are augmented images that describe the nonredundant information in the video sequence. This nonredundant information corresponds to the video sequence content. For each independently moving object in the scene, we associate a different mosaic, which we call *figure mosaic*. The "scene" can be a real 3-D scene or a synthetic scene generated by graphics tools. For the image background, which is static or slowly varying, we associate the *background mosaic*. These mosaics are stacked in layers [6], with the background mosaic at its bottom, according to how the objects in the scene move at different "depth" levels.

In GV, the object shape/velocity information is encoded by generative operators that represent video sequence content. These operators are applied to a stack of background/figure mosaics. They are (i) windowing operators; (ii) motion operators; (iii) cut-and-paste operators; and (iv) signal-processing operators. The window operators are *image window operators*, which select individual images of the sequence, or *figure window operators*, which select independently moving objects in the image; these are called *image figures*. The motion operators describe temporal transformations of window and/ or mosaics. They encode rigid translational, scaling, and rotational motion of image regions. The cut-and-paste operators are used to recursively generate mosaics from the video sequence. Finally, the signal-processing operators describe processes, e.g., spatial smoothing or scale transformation. These operators are supplemented with the Stratification Principle and the Tessellation Principle. The Stratification Principle describes how world images are stratified in layers, and the Tessellation Principle represents image figures in terms of compact geometric models.

2.3.1.1 Figure and background mosaics generation

In GV, mosaics are generated via a content-based method. This means that the image background and foreground objects are selected as whole regions by shape or windowing operators, and these regions are separately combined from frame to frame in order to generate the background and foreground mosaics.

The background mosaic Φ_B is generated recursively from the image regions selected by shape operators from consecutive images. This is realized by cut-and-paste operations. Given a sequence of N consecutive images $I_1, \cdots, I_N$, we assume that figure and camera velocities are known and that figure/background segmentation and tessellation have been completed. The background mosaic Φ_B is generated by the following recursion:

1. For $r = 1$

$$\Phi_{B,1} = M_1 I_1. \tag{2.57}$$

2. For $r \geq 2$

$$\Phi_{B,r} = A_r \Phi_{B,r-1} + B_r(M_r I_r). \tag{2.58}$$

A_r is decomposed as

$$A_r \overset{df}{=} (I - A_{2,r})A_{1,r}, \tag{2.59}$$

where I is the identity operator. $\Phi_{B,r}$ represents the world image at the recursive step r, I_r is the rth image from the sequence, and M_r is the shape operator that selects from image I_r the tessellated figure or the image background region. The operators A_r and B_r perform the operations of registration, intersection, and cutting.

The operators $A_{1,r}$ and B_r register $\Phi_{B,r}$ and I_{r-1} by using the information about camera motion. Once registered, the operator $A_{2,r}$ selects from $\Phi_{B,r-1}$ that region that it has in common with I_r, and $I - A_{2,r}$ cuts out of $\Phi_{B,r-1}$ this region of intersection. Finally, the resulting regions are pasted together. This algorithm is shown in Figure 2.8.

For example, given a sequence of 300 frames of the "Oakland" sequence taken with a handheld camera w.r.t. a static image background (see Figure 2.9 for two snapshots), the resulting mosaic is shown in Figure 2.10.

In the presence of figures, i.e., objects, we have to deal with the problem of image occlusion. Image occlusion can occur when a figure occludes another figure and/or the image background is occluded by the image boundaries. This requires the introduction of a new set of figure and background cut-and-paste operators. For conciseness, we deal here only with the case of one figure moving relative to the image background. The figure velocity is $v^F = (v_x^F, v_y^F) = (d_x^F, d_y^F)$, where d_x^F and d_y^F are integers; the image background velocity is v^I.

The figure mosaic Φ_F is generated by the recursion:

1. For $r = 1$

$$\Phi_1^B = S_1^B I_1, \tag{2.60}$$

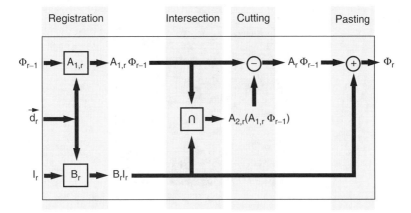

Figure 2.8 The flow diagram for the background mosaic generation algorithm. (*Source*: From R.S. Jasinschi and J.M.F. Moura, IEEE Proceedings of ICIP, 1995.)

$$\Phi_1^F = S_1^F I_1. \tag{2.61}$$

2. For $r \geq 1$

$$\Phi_r^B = (I - M_r^B)[A_{1,r}(\Phi_{r-1}^B)] + M_r^B(B_r I_r), \tag{2.62}$$

$$\Phi_r^F = O_r[A_{1,r}(\Phi_{r-1}^F)] + M_r^F(B_r I_r). \tag{2.63}$$

These expressions are structurally similar to the ones used for background mosaic generation. The only difference between (2.57) and (2.58) and (2.60), (2.61), (2.62), and (2.63) is that the latter expressions include a new set of cut-and-paste operators.

Expressions (2.60) and (2.61) compared with (2.57) contain the background S_1^B and figure S_1^F selection operators. S_1^F selects from inside image I_1 the figure region, and S_1^B selects the complement region corresponding to the unoccluded image background. S_1^B and S_1^F are instances, for $r = 1$, of the rth step background and figure selection operators S_r^B and S_r^F, respectively. S_r^B and S_r^F are $(N_x^I \cdot N_y^I) \times (N_x^I \cdot N_y^I)$ operators, i.e., they have the same dimensions that are constant because N_x^I and N_y^I are fixed.

Comparing (2.58) with (2.62), we conclude that $A_{2,r}$ is replaced by M_r^B and that $B_r I_r$ is replaced by $M_r^B(B_r I_r)$. M_r^B has the role of placing the unoccluded background region selected by S_r^B in relation to the world image coordinate system at step r. Formally, $M_r^B(S_r^B I_r)$ corresponds to the background region in image I_r placed in relation to the world image coordinate system. In the absence of figures, M_r^B reduces to $A_{2,r}$.

The figure mosaic recursion (2.63) is new. The operator M_r^F places the figure region selected by S_r^F in relation to the world image coordinate system

a.

b.

Figure 2.9 The first (a) and 100th (b) images of the "Oakland" sequence. (*Source*: From R.S. Jasinschi, J.M.F. Moura, J.C. Cheng, and A. Asif, Proceedings of IEEE ICASSP, 1995.)

at step r. Therefore, $M_r^F\left(S_r^F I_r\right)$ corresponds to the figure region in image I_r placed in relation to the mosaic coordinate system. The operator O_r will be described next.

In order to generate Φ_r^F, we have to know the operators O_r, $A_{1,r}$, and M_r^F. Among these only O_r has to be determined independently; $A_{1,r}$ and M_r^F are obtained using the information about the image window translational motion.

O_r is defined recursively. Let F_r correspond to the figure cut-and-paste operator at step r defined by

$$F_r \overset{df}{=} O_r + M_r^F. \tag{2.64}$$

Given that we know F_{r-1}, we determine F_r and O_r through the recursion:

Figure 2.10 The background mosaic for the "Oakland" sequence. It also shows the trajectory of the camera center as it moves to capture the 300 frames. (*Source*: From R.S. Jasinschi, J.M.F. Moura, J.C. Cheng, and A. Asif, Proceedings of IEEE ICASSP, 1995.)

1. Expand F_{r-1} to $B_r F_{r-1}$.
2. Perform motion compensation on $B_r F_{r-1}$. This results in the operator $K_r \overset{df}{=} [(D_V \otimes D_H)(B_r F_{r-1})]$.
3. Determine the operator L_r which computes the region of figure overlap between the ones selected by K_r and M_r^F.
4. Subtract L_r from K_r. This gives us O_r.

In summary, step 1 expands the dimensions of F_{r-1} in order to match it to that of the figure mosaic at step r. In step 2 we translate the figure part selected from image I_{r-1} to the position it should have at step r; this is realized by premultiplying $B_r F_{r-1}$ with the dislocation operator; the powers of the row and column component dislocation operators are equal to $d_x^F + d_x^I$ for D_V, and $d_y^F + d_y^I$ for D_H; we have to compensate for both image window and figure translational motion. In step 3, we determine the operator, which selects the figure part that is in overlap between the figure selected from image I_r and I_{r-1}, properly compensated and dimensionally scaled. Finally, in step 4 we determine the figure part that is only known at step $r - 1$, i.e., the region selected by O_r.

As an example of figure and background mosaic generation, we use a sequence, called the "Samir" sequence, of thirty 240×256 images of a real 3-D scene recorded through a handheld camera. The sequence shows a

a.

b.

Figure 2.11 The first (a) and 30th (b) images of the "Samir" sequence. (*Source*: From R.S. Jasinschi and J.M.F. Moura, IEEE Proceedings of ICIP, 1995.)

person walking in front of a building, the Wean Hall, at the Carnegie Mellon University campus. The illumination conditions were that of a sunny bright day. Some images are shown in Figure 2.11.

First, we segment the figure in relation to the image background by detecting the image region with highest velocity. We use a detection measure on a Gaussian pyramid of the sequence, followed by thresholding. For each pair of images, we obtain a binary image showing the figure shape. Since the Samir figure moves nonrigidly, we obtain a different segmented figure for each image pair. Second, we tessellate the segmented figure. We tessellate the person's torso and head separately from his feet and legs because his head and torso move rigidly while his feet and legs move nonrigidly. As a result of this, we obtain a tessellated figure that is the union of the tessellated head and torso with the tessellated feet and legs. We determine the velocity of this tessellated figure through rectangle matching. This tessellated figure

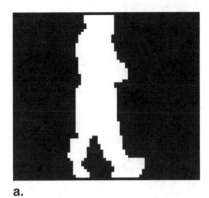

a.

b.

Figure 2.12 The motion-based segmented and tessellated "Samir" sequence for the 38th frame: (a) the motion-based segmentation result; (b) the tessellation result. (*Source*: From R.S. Jasinschi and J.M.F. Moura, IEEE Proceedings of ICIP, 1995.)

with its velocity corresponds to the figure and motion operators. Figure 2.12 shows the motion-based segmented and tessellated figure of the person for the 22nd image.

In Figure 2.13, we show the background mosaic as processed in the 5th and 28th frames; the latter represents the final background mosaic. The foreground mosaic is shown in Figure 2.14.

2.3.2   3-D based mosaics

In this method, partial or full 3-D information, e.g., object depth or velocity, is used to generate mosaics. This generalizes GV by using this extra 3-D information for mosaic generation. At the core of this method [65, 66, 67] lies the detailed extraction of 3-D information, i.e., structure from motion.

a.

b.

Figure 2.13 The background mosaic at the 5th (a) and 11th (b) frames of the "Samir" sequence. (*Source*: From R.S. Jasinschi and J.M.F. Moura, IEEE Proceedings of ICIP, 1995.)

The standard structure from motion methods are designed to extract the 3-D object shape (via its depth) and velocity information. This requires the selection of points on these objects, which are tracked in time via their 2-D associated image velocities, that introduces an ad hoc factor: the (3-D) object shape is computed by using the *a priori* knowledge of its associated (2-D) shape. We solved this "chicken and egg" problem by generalizing the known 8-point algorithm [68, 69, 70], as described below. Based on this method, a dense 3-D scene depth map is generated. This depth map corresponds to the background part of the image, i.e., the static or slowly varying part of the 3-D scene. In addition to this, the camera (3-D) velocity is computed. Foreground (independently from the camera and background) moving objects are selected. Finally, based on this 3-D, as well as 2-D, information, a set of layered mosaics is generated. We distinguish between planar mosaics and 3-D mosaics. The former are recursively generated from parts of 2-D

a.

b.

Figure 2.14 The foreground (a) and background (b) mosaics for the "Samir" sequence. (*Source*: From R.S. Jasinschi and J.M.F. Moura, IEEE Proceedings of ICIP, 1995.)

images, whereas the latter uses full 3-D depth information. These elements are described next.

2.3.2.1   Structure from motion: generalized eight-point algorithm
The eight-point algorithm for the computation of scene structure (shape/depth) and (camera/object) motion (8PSFM) as introduced by Longuett-Higgins [68] and further developed by Tsai and Huang [69] is simple. Given two views from an uncalibrated camera of a rigidly moving object in a 3-D scene, by using the image correspondence of eight (or more) points projected from the scene onto the two view images, the structure and motion of this object can be computed. Given these eight points, 8PSFM computes the fundamental matrix; this matrix combines the rigid body assumption, the perspective projection of points in 3-D space onto the

image plane, and the correspondence of points between two successive images. Using the essential matrix, the 3-D object rotation and translation matrices are computed, and from these the relative depths of points on the object are estimated. The limitations of 8PSFM are its sensitivity to noise (SVD matrix computation) and the choice of points on the object. We proposed [66] a generalization of 8PSFM, called G8PSFM, that avoids the need of choosing in an ad hoc manner (image projected) object points, making the computation of SFM more robust.

The main elements of G8PSFM are:

1. Divide two successive images into eight approximately identical rectangular blocks; each block contains approximately the same number of feature points, and each point belongs to one and only one block.

2. From each block in each image, randomly draw feature points according to a uniformly distributed random number generator. The result of this is a set of eight feature point correspondences that spans over the whole of the two images; the coordinates of the feature points are normalized, so that $(0,0)$ is at the center of the image plane, and the width and height of the image are each 2.

3. Track corresponding points based on a cross-correlation measure and bidirectional consistency check; two pruning methods reduce the population of tracked features to a set of stable points. First, all feature points with poor cross-correlation (based on a threshold) are eliminated. Second, only the feature points that satisfy a bidirectionality criterion are retained.

4. Compute the fundamental matrix.

5. Compute the three rotation matrix R components and the two translation matrix T components.

6. Compute the depth of the corresponding points in the 3-D scene.

The above steps are repeated enough times to exhaust all possible combinations of feature points. This approach can become computationally intractable, thus making the total number too large for real implementations. We devised a robust suboptimum statistical approach that determines the choice of the "best" set of camera motion parameters from a subset of all possible combinations. This uses an overall velocity estimation error as the quality measure (see [66]).

We implemented G8PSFM by assuming that we presegment independently moving foreground objects in the scene. Thus, for the purpose of layered mosaic generation, we process only the static part of the scene. The G8PSFM method can be used in the general case when we have an arbitrary number of objects moving in the scene. In the discussion that follows, the static background is identified as a single rigid object. We assume a pinhole camera with unit focal length. The camera moves in the 3-D scene, inducing image motion, e.g., tracking or booming. The camera velocity w.r.t. this background is the negative of the velocity of the background w.r.t. the camera.

Figure 2.15 Results for the "Flowergarden" sequence using the G8PSFM method. (*Source*: From R.S. Jasinschi, T. Naveen, A.J. Tabatabai, and P. Babic-Vovk, IEEE Transactions on Circuits and Systems for Video Technology, 10(7): October 2000.)

After all these operations, a dense depth map, i.e., depth values at all image points, is computed. For this a Delaunay triangulation of the feature points is performed, and a linear interpolation of the depth values available at the vertices of triangles is computed, thus filling in all internal triangle points with depth data.

Figure 2.15 shows the results of applying the G8PSFM to the "Flowergarden" sequence. The top image shows an original (gray-level) image; the middle image displays the dense depth map; the lower image shows how, e.g., the tree is segmented from the top image by using depth map information from the middle image.

2.3.2.2   *Layered mosaics based on 3-D information*

Mosaic generation based on 3-D information differs from traditional methods (photo-mosaics) that rely on strictly 2-D image information, e.g., color, motion, and texture, by using 3-D depth information and camera motion parameters. Thus, the method discussed here [66, 67] generalizes the photo-mosaic generation methods in that it:

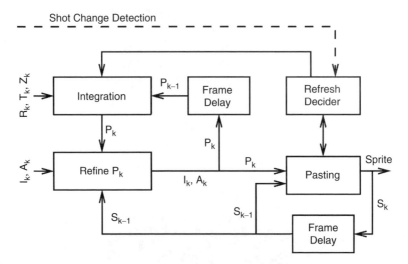

Figure 2.16 Overall flow diagram for the layered mosaic generation method based on 3-D information. (*Source*: From R.S. Jasinschi, T. Naveen, A.J. Tabatabai, and P. Babic-Vovk, IEEE Transactions on Circuits and Systems for Video Technology, 10(7): October 2000.)

1. Computes explicitly 3-D camera parameters.
2. Determines the structure (depth) of static scenes.
3. Integrates camera and depth information with a perspective transformation model of image point displacements.
4. Models explicitly variations in image illumination.
5. Can deal with arbitrary camera velocities in that it is described by an image multiresolution method.
6. Can be extended to deal with 3-D surfaces.

In our approach shown in Figure 2.16, we identify regions or layers of "uniform" depth in the scene, and we generate a 2-D sprite for each such layer. It is assumed that:

1. Independently moving foreground objects are presegmented.
2. The segmentation of the background based on depth has already taken place.
3. Based on the first two assumptions, the portion of background for which a sprite is being created is determined by an α-map A_k.

The main elements of the method are shown in Figure 2.16 and a detailed description is given in [66]. In general, a mosaic M_k at time k is generated by integrating images of a video sequence from time 1 through k; this can apply to the luma (Y video component) or to chroma (Cr and Cb video components). In this process, the mosaic M_k is obtained by using a prediction using a nine-parameter plane perspective projection $M_k(x,y) = M_k(f_k(x,y), g_k(x,y))$, where

$$f_k(x,y) = \frac{P_k(0,0)\cdot x + P_k(0,1)\cdot y + P_k(0,2)}{P_k(2,0)\cdot x + P_k(2,1)\cdot y + P_k(2,2)}$$

$$g_k(x,y) = \frac{P_k(1,0)\cdot x + P_k(1,1)\cdot y + P_k(1,2)}{P_k(2,0)\cdot x + P_k(2,1)\cdot y + P_k(2,2)},$$

(2.65)

and $P_k(\cdot,\cdot)$ is a 3×3 matrix, with $P_k(2,2)$ equal to 1. The goal then is to estimate the eight components of the matrix P_k such that $\hat{I}_k \approx I_k, \forall k$. It is done as follows. As an initialization of the process, the first sprite generated is equal to the first input image ($M_1 = I_1$) and P_1 an identity matrix. Next, for $k > 1$, the 3-D G8PSFM parameters are integrated with $P_{k-1}(\cdot,\cdot)$ to obtain P_k, as described in the following. This is done by mapping estimated 3-D parameters R_k, T_k, and Z_k into P_k. The 3-D parameter estimation was performed on a pair of pictures I_{k-1} and I_k. Consequently, the estimated camera parameters are relative to the camera position at instant $k - 1$. However, since the parameters P_k are relative to the mosaic, a mechanism is needed to go from relative (i.e., pairwise) 3-D parameters to relative 2-D parameters and to accumulate these relative 2-D parameters over time to form absolute 2-D parameters. Key to this is the computation of the 3×3 Q matrix. Given a point (x_n, y_n) in normalized coordinates of image I_k, Q allows us to find an approximate location (x'_n, y'_n) of the corresponding point in image I_{k-1} as:

$$x'_n = \frac{Q(0,0)\cdot x_n + Q(0,1)\cdot y_n + Q(0,2)}{Q(2,0)\cdot x_n + Q(2,1)\cdot y_n + Q(2,2)}$$

$$y'_n = \frac{Q(1,0)\cdot x_n + Q(1,1)\cdot y_n + Q(1,2)}{Q(2,0)\cdot x_n + Q(2,1)\cdot y_n + Q(2,2)}.$$

(2.66)

The mapping from pairwise 3-D to pairwise 2-D parameters Q is performed by assuming that the particular region in 3-D for which a mosaic is being generated is a 2-D planar object with approximately uniform depth. This representative depth $\tilde{Z}$ for relevant objects in I_{k-1} is obtained as:

$$\tilde{Z} = \text{median}\left\{ Z_{k-1}(x,y) \middle| A_{k-1}(x,y) \geq \tau_a \right.$$

(2.67)

where τ_a is a threshold determined empirically. The image

$$\left\{(x,y): A_{k-1}(x,y) \geq \tau_a\right\}$$

is usually referred to as the α-image in the MPEG IV community. Then, Q can be computed as

$$Q \triangleq R_{k-1}^{-1} - \frac{1}{\tilde{Z}} R_{k-1}^{-1} T[0\ 0\ 1], \tag{2.68}$$

where R_{k-1} and T_{k-1} are the rotation and translation matrices in 3-D, respectively. The entry $Q(2,2)$ equals 1.0. We arrived at these relations based on successive approximations and simplifications made to a generic situation with 3-D scenes and a real camera to 2-D planar objects at uniform depth and a pinhole camera, for which $Z = \tilde{Z}$. The final step of combining the Q and P matrices' parameters is described in [65].

As a consequence of this method, for each 3-D region for which we can define a consistent average depth $\tilde{Z}$ that represents the average depth of the layer associated to this region, we generate a separate mosaic. Thus, e.g., for the "Flowergarden" sequence we should expect to have at least three layers: the foreground tree, the flowerbed, and the houses/sky. This is a crude approximation because the flowerbed itself is represented by a receding plane on which, upon lateral camera motion, different points move at different speeds; the higher the speed, the closer they are to the camera and vice versa. In Figure 2.17 and Figure 2.18 we show the mosaics for the flowerbed and houses obtained with 150 images.

Figure 2.17 The "flowerbed" layered mosaic for the "Flowergarden" sequence using 150 frames. (*Source*: From R.S. Jasinschi, T. Naveen, A.J. Tabatabai, and P. Babic-Vovk, IEEE Transactions on Circuits and Systems for Video Technology, 10(7): October 2000.)

Figure 2.18 The "houses" layered mosaic for the "Flowergarden" sequence using 150 frames. (*Source*: From R.S. Jasinschi, T. Naveen, A.J. Tabatabai, and P. Babic-Vovk, IEEE Transactions on Circuits and Systems for Video Technology, 10(7): October 2000.)

2.3.2.3 3-D mosaics

3-D mosaics are generated by performing photometric and depth map compositing. The difference between full 3-D mosaics and layered 3-D-based mosaics is that in the latter case depth approximations are used in the form of discrete depth layers, whereas for the former case depth is full. Also, depth maps have to be composited for 3-D mosaics. This introduces a new source of uncertainty: the incremental composition of depth maps. We have a partial solution to this problem that is realized by depth map registration and depth map composition. Depth map registration is shown in Figure 2.19.

Depth map compositing is realized recursively. Given the video sequence $\{I_1, \cdots, I_k, \cdots, I_N\}$, we want to generate an extended depth map by combining the depth maps generated between pairs of successive images.

For each pair of successive images, we generate two depth maps, i.e., $\{Z_k^F\}$ and $\{Z_{k+1}^P\}$ ("F" stands for *future* and "P" stands for *past*). The important thing to notice is that both depth maps share the same scaling factor as defined in structure from motion. They can be combined without creating any ambiguity[3]. The combination of these depth maps is described by the operation $\oplus : \{Z_k^F\}, \{Z_{k+1}^P\} \rightarrow \{Z_k^F \oplus Z_{k+1}^{P,R}\}$. We observe that $Z_{k+1}^{P,R}$ has the superscript R, which denotes the fact that the depth map $\{Z_{k+1}^P\}$ has been registered to $\{Z_k^F\}$ by using the translation and rotation parameters estimated between images I_k and I_{k+1}. The extended depth map $\{Z_k^F \oplus Z_{k+1}^{P,R}\}$ includes each individual depth map; the operation $\oplus$ represents each 3-D point with a single depth value; multiple values at the same point are averaged out. Since as the camera moves, e.g., from I_k to I_{k+1}, we see new parts of the 3-D scene in I_{k+1}, which were not visible in I_k, and vice versa. $\{Z_k^F \oplus Z_{k+1}^{P,R}\}$ contains more depth points than in each individual depth map.

The first instance of the extended depth map is given by combining the depth maps generated between images I_1 and I_2, thus resulting in $\{Z_1^F \oplus Z_2^{P,R}\}$. The process of depth map composition is recursive and involves two processing layers, which describe the combination of pairs and multiple depth maps. Formally:

1. For each pair of images, I_k and I_{k+1}, generate pairs of depth maps $\{Z_k^F\}$ and $\{Z_{k+1}^P\}$. This uses the result of structure from motion.
2. Generate, for each pair of images, an extended depth map $\{Z_k^F \oplus Z_{k+1}^{P,R}\}$. This is realized through the compositing operation $\oplus$ and depth map registration denoted by R.
3. Compose $\{Z_k^F \oplus Z_{k+1}^{P,R}\}$ with the extended depth map obtained up to image I_k. This involves scale equalization, denoted by $\div$ and depth map registration.

---

[3] If we combine depth maps with different scaling factors or without equalized scaling factors, the result is meaningless because depth values for the same 3-D point are different.

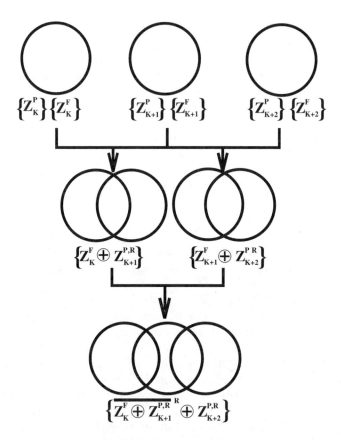

Figure 2.19   Depth map generation block diagram. Each circle represents a depth map. Using the depth map generated among consecutive images, i.e., I_k, I_{k+1}, and I_{k+2}, a composited depth map, shown at the bottom layer, is generated in two steps. Between the top layer and the middle layer a composited depth map is generated from pairs of depth maps generated between pairs of successive images. Finally, between the middle layer and the bottom layer, a single depth map is generated by combining the previous two depth maps.

Scale equalization is the operation by which depth values are rescaled in order to adjust them to the same scale. Only the pairs of depth maps $\{Z_k^F\}$ and $\{Z_{k+1}^P\}$, generated for successive images I_k and I_{k+1} using structure from motion by using the same translation and rotation parameters, have the same scale; depth maps obtained for different successive pairs of images depend on different scaling values.

The depth compositing described above is repeated until all images have been processed.

Figure 2.20, Figure 2.21, and Figure 2.22 display the recursive process of depth map composition. Given three consecutive images I_k, I_{k+1}, and I_{k+2}, we first obtain pairs of (equiscalar) depth maps $\{Z_k^F\}/\{Z_{k+1}^P\}$ and $\{Z_{k+1}^F\}/\{Z_{k+2}^P\}$. The result is displayed by circles representing a depth map pair.

Figure 2.20 Frontal view of the VRML rendering of a 3-D mosaic.

Figure 2.21 Top view of the VRML rendering of a 3-D mosaic. We can observe that the tree protrudes in the depth map because it lies at a smaller depth w.r.t. to the camera compared to the other parts of the scene.

Next, we combine pairs of successive depth maps, resulting in the extended depth maps $\{Z_k^F \oplus Z_{k+1}^{P,R}\}$ and $\{Z_{k+1}^F \oplus Z_{k+2}^{P,R}\}$; they are represented by intersecting circles. Finally, the extended pairs of depth maps are integrated into a single depth map $\overline{\{Z_k^F \oplus Z_{k+1}^{P,R}\}}^R \oplus Z_{k+2}^{P,R}$.

Figures 2.20, 2.21, and 2.22 show the VRML rendering of a photometric and depth map compositing. This was realized by associating voxels (3-D volumetric units) to different depth values and doing the texture mapping (photometric map).

2.3.2.4 Summary

We showed a process of evolution and complexification of mosaic generation: from 2-D to 3-D mosaics. We described generative video, a much richer type of mosaic than traditional 2-D mosaics. The more 3-D information is incorporated into this process of mosaic generation, the more complete the

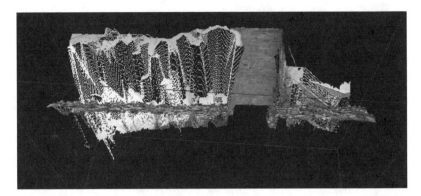

Figure 2.22 Bottom view of the VRML rendering of a 3-D mosaic. The flowerbed is aligned perpendicular to the field of view. We can also observe gaps (black holes) in the 3-D mosaic because from this viewing angle the voxels are not aligned with the viewing angle.

mosaic becomes in terms of its fidelity to full 3-D information. On the other hand, more levels of uncertainty and imperfections are added in the process. The final goal of having a full 3-D reconstruction of a 3-D scene with precise depth and photometric compositing is still much beyond reach, but this goal is seen as a driving element in the process described in this section. In the next section, we describe a full 3-D representation of objects that is extracted from monocular video sequences.

2.4   Three-dimensional object-based representation

In this section we describe a framework for 3-D model-based digital video representation. The proposed framework represents a video sequence in terms of a 3-D scene model and a sequence of 3-D pose vectors. The 3-D model for the scene structure contains information about the 3-D shape and texture. The 3-D shape is modeled by a piecewise planar surface. The scene texture is coded as a set of ordinary images, one for each planar surface patch. The pose vectors represent the position of the camera with respect to the scene. A shorter version of the material in this section was presented in [71].

The main task in analyzing a video sequence within our framework is the automatic generation of the 3-D models from a single monocular video data sequence. We start by reviewing in Subsection 2.4.1 the existing methods to recover 3-D structure from video. Then, we describe the proposed framework in Subsection 2.4.2 and detail the analysis and synthesis tasks in subsections 2.4.3 and 2.4.4. Finally, in Subsections 2.4.5 and 2.4.6, we describe experiments and applications of the proposed framework.

2.4.1   3-D object modeling from video

For videos of unconstrained real-world scenes, the strongest cue to estimating the 3-D structure is the 2-D motion of the brightness pattern in the image plane; thus, the problem is generally referred to as SFM. The two major steps in SFM are usually the following: first, compute the 2-D motion in the image plane; second, estimate the 3-D shape and the 3-D motion from the computed 2-D motion.

Early approaches to SFM processed a single pair of consecutive frames and provided existence and uniqueness results to the problem of estimating 3-D motion and absolute depth from the 2-D motion in the camera plane between two frames; see, for example, [69]. The two-frame algorithms are highly sensitive to image noise and, when the object is far from the camera, i.e., at a large distance when compared to the object depth, they fail even at low-level image noise. More recent research has been oriented toward the use of longer image sequences. For example, [72] uses nonlinear optimization to solve for the rigid 3-D motion and the set of 3-D positions of feature points tracked along a set of frames, and [73] uses a Kalman filter to integrate along time a set of two-frame depth estimates.

Among the existing approaches to the multiframe SFM problem, the factorization method, introduced by Tomasi and Kanade [12], is an elegant method to recover structure from motion without computing the absolute depth as an intermediate step. They treat orthographic projections. The object shape is represented by the 3-D position of a set of feature points. The 2-D projection of each feature point is tracked along the image sequence. The 3-D shape and motion are then estimated by factorizing a measurement matrix whose entries are the set of trajectories of the feature point projections. Tomasi and Kanade pioneered the use of linear subspace constraints in motion analysis. In fact, the key idea underlying the factorization method is the fact that the rigidity of the scene imposes that the measurement matrix lives in a low-dimensional subspace of the universe of matrices. Tomasi and Kanade have shown that the measurement matrix is a rank 3 matrix in a noiseless situation. This work was later extended to the scaled-orthographic — pseudoperspective — and paraperspective projections [74], correspondences among line segments [75], recursive formulation [76], and multibody scenario [77].

2.4.1.1   Surface-based rank 1 factorization method

We use linear subspace constraints to solve SFM, recovering 3-D motion and a parameteric description of the 3-D shape from a sequence of 2-D motion parameters. By exploiting the subspace constraints, we solve the SFM problem by factorizing a matrix that is rank 1 in a noiseless situation rather than a rank 3 matrix as in the original factorization method.

To recover in an expedient way the 3-D motion and the 3-D shape, we develop the surface-based rank 1 factorization method [13]. Under our general scenario, we describe the shape of the object by surface patches. Each patch is described by a polynomial. This leads to a parameterization of the object surface. We show that this parametric description of the 3-D shape induces a parametric model for the 2-D motion of the brightness pattern in the image plane. The surface-based factorization approach [13] overcomes a limitation of the original factorization method of Tomasi and Kanade [12]. Their approach relies on the matching of a set of features along the image sequence. To provide dense depth estimates, their method usually needs hundreds of features that are difficult to track and that lead to a complex correspondence problem. Instead of tracking pointwise features, the surface-based method tracks regions where the optical flow is described by a single set of parameters. This approach avoids the correspondence problem and is particularly suited to practical scenarios such as when constructing 3-D models for buildings that are well described by piecewise flat surfaces.

The algorithm that we develop has a second major feature — its computational simplicity. By making an appropriate linear subspace projection, we show that the unknown 3-D structure can be found by factorizing a matrix that is rank 1 in a noiseless situation [14]. This contrasts with the factorization of a rank 3 matrix as in the original method of Tomasi and Kanade [12]. This allows the use of faster iterative algorithms to compute the matrix that best approximates the data.

Our approach handles general-shaped structures. It is particularly well suited to the analysis of scenes with piecewise flat surfaces, where the optical flow model reduces to the well known affine motion model. This is precisely the kind of scenes for which the piecewise mosaics video representation framework is particularly suited. References [13, 14] contain the detailed theoretical foundations of our approach to video analysis. In this section, we particularize to the construction of piecewise mosaics our general methodology.

2.4.2   Framework

We now detail the framework for 3-D model-based video representation. We start by describing the video model as a sequence of projections of a 3-D scene. Then we consider the representation of the 3-D motion of the camera and the 3-D shape of the scene.

2.4.2.1   Image sequence representation

For commodity, we consider a rigid object $\mathcal{O}$ moving in front of a camera. The object $\mathcal{O}$ is described by its 3-D shape S and texture $\mathcal{T}$. The texture $\mathcal{T}$ represents the light received by the camera after reflecting on the object surface, i.e., the texture $\mathcal{T}$ is the object brightness as perceived by the camera. The texture depends on the object surface photometric properties, as well as on the environment illumination conditions. We assume that the texture does not change with time.

The 3-D shape S is a representation of the surface of the object, as detailed below. The position and orientation of the object $\mathcal{O}$ at time instant f is represented by a vector m_f. The 3-D structure obtained by applying the 3-D rigid transformation coded by the vector m_f to the object $\mathcal{O}$ is represented by $\mathcal{M}(m_f)\mathcal{O}$.

The frame I_f, captured at time f, is modeled as the projection of the object

$$I_f = \mathcal{P}\{\mathcal{M}(m_f)\mathcal{O}\}. \tag{2.69}$$

We assume that $\mathcal{P}$ is the orthogonal projection operator that is known to be a good approximation to the perspective projection when the relative depth of the scene is small in relation to the distance to the camera. Our video analysis algorithms can be easily extended to the scaled-orthography and the paraperspective models in a similar way that [74] does for the original factorization method.

The operator $\mathcal{P}$ returns the texture $\mathcal{T}$ as a real-valued function defined over the image plane. This function is a nonlinear mapping that depends on the object shape S and the object position m_f. The intensity level of the projection of the object at pixel u on the image plane is

$$\mathcal{P}\{\mathcal{M}(m_f)O\}(u) = \mathcal{T}\big(s_f(S, m_f; u)\big), \tag{2.70}$$

where $s_f(S, m_f; u)$ is the nonlinear mapping that lifts the point u on the image I_f to the corresponding point on the 3-D object surface. This mapping $s_f(S, m_f; u)$ is determined by the object shape S and the position m_f. To simplify the notation, we will usually write explicitly only the dependence on f, i.e., $s_f(u)$.

The image sequence model (2.69) is rewritten in terms of the object texture $\mathcal{T}$ and the mappings $s_f(u)$, by using the equality (2.70), as

$$I_f(u) = \mathcal{T}\big(s_f(u)\big). \tag{2.71}$$

Again, the dependence of s_f on S and m_f is omitted for simplicity.

2.4.2.2 3-D motion representation

To represent the 3-D motion, we attach coordinate systems to the object and to the camera. The object coordinate system (o.c.s.) has axes labeled by x, y, and z, while the camera coordinate system (c.c.s.) has axes labeled by u, v, and w. The plane defined by the axes u and v is the camera plane. The unconstrained 3-D motion of a rigid body can be described in terms of a time-varying point translation and a rotation (see [78]).

The 3-D motion of the object is then defined by specifying the position of the o.c.s. $\{x, y, z\}$ relative to the c.c.s. $\{u, v, w\}$, i.e., by specifying a rotation–translation pair that takes values in the group of the rigid transformations of the space, the special Euclidean group SE(3). We express the object position at time instant f in terms of $\left(t_f, \boldsymbol{\Theta}_f\right)$ where the vector $t_f = \left[t_{uf}, t_{vf}, t_{wf}\right]^T$ contains the coordinates of the origin of the object coordinate system with respect to the camera coordinate system (translational component of the 3-D motion), and $\boldsymbol{\Theta}_f$ is the rotation matrix that represents the orientation of the object coordinate system relative to the camera coordinate system (rotational component of the 3-D motion).

2.4.2.3   3-D shape representation

The 3-D shape of the rigid object is a parametric description of the object surface. We consider objects whose shape is given by a piecewise planar surface with K patches. The 3-D shape is described in terms of K sets of parameters $\left\{a_{00}^k, a_{10}^k, a_{01}^k\right\}$, for $1 \le k \le K$, where

$$z = a_{00}^k + a_{10}^k(x - x_0^k) + a_{01}^k(y - y_0^k) \tag{2.72}$$

describes the shape of the patch k in the o.c.s. With respect to the representation of the planar patches, the parameters x_0^k and y_0^k can have any value; for example they can be made zero. We allow the specification of general parameters $\left(x_0^k, y_0^k\right)$ because the shape of a small patch k with support region $\{(x, y)\}$ located far from the point $\left(x_0^k, y_0^k\right)$ has a high sensitivity with respect to the shape parameters. To minimize this sensitivity, we choose for $\left(x_0^k, y_0^k\right)$ the centroid of the support region of patch k. With this choice, we improve the accuracy of the 3-D structure recovery algorithm.

By making $\left(x_0^k, y_0^k\right)$ the centroid of the support region of patch k, we also improve the numerical stability of the algorithm that estimates the 2-D motion in the image plane.

The piecewise planar 3-D shape described by Expression (2.72) captures also the simpler feature-based shape description. This description is obtained by making zero all the shape parameters, except for a_{00}^k that codes the relative depth of feature k, $z = a_{00}^k$.

2.4.3   Video analysis

The video analysis task consists in recovering the object shape, object texture, and object motion from the given video. This task corresponds to inverting the relation expressed in Equation (2.71); i.e., we want to infer the 3-D shape S, the texture $\mathcal{T}$, and the 3-D motion $\left\{m_f, 1 \le f \le F\right\}$ of the object $\mathcal{O}$ from the video sequence $\left\{I_f, 1 \le f \le F\right\}$ of F frames. In [13] we study this problem for a piecewise polynomial shape model. This section particularizes our approach for piecewise planar shapes.

We infer the 3-D rigid structure from the 2-D motion induced onto the image plane. After recovering the 3-D shape and the 3-D motion of the object, the texture of the object is estimated by averaging the video frames coregistered according to the recovered 3-D structure. We now detail each of these steps.

2.4.3.1 Image motion

The parametric description of the 3-D shape induces a parameterization for the 2-D motion in the image plane $\{u_f(s)\}$. The displacement between the frames I_1 and I_f in the region corresponding to surface patch k is expressed in terms of a set of 2-D motion parameters. For planar patches, we get the affine motion model for the 2-D motion of the brightness pattern in the image plane. We choose the coordinate $s = [s, r]^T$ of the generic point in the object surface to coincide with the coordinates $[x,y]^T$ of the object coordinate system. We also choose the object coordinate system so that it coincides with the camera coordinate system in the first frame ($t_{u1} = t_{v1} = 0, \boldsymbol{\Theta}_1 = I_{3\times3}$). With these definitions, we show in [13] that the 2-D motion between the frames I_1 and I_f in the region corresponding to surface patch k is written as

$$u_f(s) = u(\boldsymbol{\alpha}_f^k; s, r) = \begin{bmatrix} \alpha_{f10}^{uk} & \alpha_{f01}^{uk} \\ \alpha_{f10}^{vk} & \alpha_{f01}^{vk} \end{bmatrix} \begin{bmatrix} s - x_0^k \\ r - y_0^k \end{bmatrix} + \begin{bmatrix} \alpha_{f00}^{uk} \\ \alpha_{f00}^{vk} \end{bmatrix}, \tag{2.73}$$

where the 2-D motion parameters $\boldsymbol{\alpha}_f^k = \left\{ \alpha_{f00}^{uk}, \alpha_{f10}^{uk}, \alpha_{f01}^{uk}, \alpha_{f00}^{vk}, \alpha_{f10}^{vk}, \alpha_{f01}^{vk} \right\}$ are related to the 3-D shape and 3-D motion parameters by

$$\alpha_{f00}^{uk} = t_{uf} + i_{xf} x_0^k + i_{yf} y_0^k + i_{zf} a_{00}^k, \tag{2.74}$$

$$\alpha_{f10}^{uk} = i_{xf} + i_{zf} a_{10}^k, \tag{2.75}$$

$$\alpha_{f01}^{uk} = i_{yf} + i_{zf} a_{01}^k, \tag{2.76}$$

$$\alpha_{f00}^{vk} = t_{vf} + j_{xf} x_0^k + j_{yf} y_0^k + j_{zf} a_{00}^k, \tag{2.77}$$

$$\alpha_{f10}^{vk} = j_{xf} + j_{zf} a_{10}^k, \tag{2.78}$$

$$\alpha_{f01}^{vk} = j_{yf} + j_{zf} a_{01}^k, \tag{2.79}$$

where $i_{xf}, i_{yf}, i_{zf}, j_{xf}, j_{yf},$ and j_{zf} are entries of the well known 3-D rotation matrix $\boldsymbol{\Theta}_f$ (see [78]).

The parameterization of the 2-D motion, as in Expression (2.73), is the well-known affine motion model. Because the object coordinate system coincides with the camera coordinate system in the first frame, the surface patch k is projected on frame 1 into region $\mathcal{R}_k$. The problem of estimating the support regions by segmenting the 2-D motion has been addressed in the past (see, for example, [79, 80]). We use the simple method of sliding a rectangular window across the image and detect abrupt changes in the 2-D motion parameters. We use known techniques to estimate the 2-D motion parameters (see [81]). Another possible way to use our structure from motion approach is to select *a priori* the support regions $\mathcal{R}_k$, as it is usually done by the feature-tracking approach. In fact, as mentioned above, our framework is general enough to accommodate the feature-based approach because it corresponds to selecting *a priori* a set of small (pointwise) support regions $\mathcal{R}_k$ with shape described by z = constant in each region. In [14] we exploit the feature-based approach.

2.4.3.2   3-D structure from 2-D motion

The 2-D motion parameters are related to the 3-D shape and 3-D motion parameters through Expressions (2.74–2.79). These expressions define an overconstrained equation system with respect to the 3-D shape parameters $\{a_{00}^k, a_{10}^k, a_{01}^k, 1 \le k \le K\}$ and to the 3-D positions

$$\left\{ m_f = \left\{ t_{uf}, t_{vf}, \boldsymbol{\theta}_f \right\}, 1 \le f \le F \right\}$$

(under orthography, the component of the translation along the camera axis, t_{wf}, cannot be recovered). The estimate $\{\hat{a}_{mn}^k\}$ of the object shape and the estimate $\left\{ \hat{t}_{uf}, \hat{t}_{vf}, \hat{\boldsymbol{\Theta}}_f \right\}$ of the object positions are the least squares (LS) solution of the system. Our approach to this nonlinear LS problem is the following. First, solve for the translation parameters, which leads to a closed-form solution. Then, replace the estimates $\left\{ \hat{t}_{uf}, \hat{t}_{vf} \right\}$ and solve for the remaining motion parameters $\{\boldsymbol{\Theta}_f\}$ and shape parameters $\{a_{mn}^k\}$ by using a factorization approach.

2.4.3.3   Translation estimation

The translation components along the camera plane at instant f, t_{uf} and t_{vf}, affect only the set of parameters $\{\alpha_{f00}^{uk}, 1 \le k \le K\}$ and $\{\alpha_{f00}^{vk}, 1 \le k \le K\}$, respectively. If the parameters $\{a_{00}^k\}$ and $\{\boldsymbol{\Theta}_f\}$ are known, to estimate $\{t_{uf}, t_{vf}\}$ is a linear LS problem. Without loss of generality, we choose the object coordinate system in such a way that $\sum_{k=1}^{K} a_{00}^k = 0$ and the image origin in such a way that $\sum_{k=1}^{K} x_0^k = \sum_{k=1}^{K} y_0^k = 0$. With this choice, the estimates $\hat{t}_{uf}$ and $\hat{t}_{vf}$ of the translation along the camera plane at frame f are

$$\hat{t}_{uf} = \frac{1}{K} \sum_{k=1}^{K} \alpha_{f00}^{uk} \quad \text{and} \quad \hat{t}_{vf} = \frac{1}{K} \sum_{k=1}^{K} \alpha_{f00}^{vk}. \tag{2.80}$$

2.4.3.4 Matrix of 2-D motion parameters

Replace the set of translation estimates $\{\hat{t}_{uf}$ and $\hat{t}_{vf}\}$ in the equation set (2.74–2.79). Define a set of parameters $\{\beta_f^{uk}, \beta_f^{vk}\}$ related to $\{\alpha_{f00}^{uk}, \alpha_{f00}^{vk}\}$ by

$$\beta_f^{uk} = \alpha_{f00}^{uk} - \frac{1}{K} \sum_{l=1}^{K} \alpha_{f00}^{ul} \quad \text{and} \quad \beta_f^{vk} = \alpha_{f00}^{vk} - \frac{1}{K} \sum_{l=1}^{K} \alpha_{f00}^{vl}. \tag{2.81}$$

Collect the parameters $\{\beta_f^{uk}, \alpha_{f01}^{uk}, \alpha_{f10}^{uk}, \beta_f^{vk}, \alpha_{f01}^{vk}, \alpha_{f10}^{vk}, 2 \le f \le F, 1 \le k \le K\}$ in a $2(F-1) \times 3K$ matrix R, which we call the matrix of 2-D motion parameters,

$$R = \begin{bmatrix} \beta_2^{u1} \alpha_{210}^{u1} \alpha_{201}^{u1} \beta_2^{u2} \alpha_{210}^{u2} \alpha_{201}^{u2} \cdots \beta_2^{uK} \alpha_{210}^{uK} \alpha_{201}^{uK} \\ \beta_3^{u1} \alpha_{310}^{u1} \alpha_{301}^{u1} \beta_3^{u2} \alpha_{310}^{u2} \alpha_{301}^{u2} \cdots \beta_3^{uK} \alpha_{310}^{uK} \alpha_{301}^{uK} \\ \cdots \cdots \cdots \cdots \cdots \cdots \cdots \cdots \cdots \\ \beta_F^{u1} \alpha_{F10}^{u1} \alpha_{F01}^{u1} \beta_F^{u2} \alpha_{F10}^{u2} \alpha_{F01}^{u2} \cdots \beta_F^{uK} \alpha_{F10}^{uK} \alpha_{F01}^{uK} \\ \beta_2^{v1} \alpha_{210}^{v1} \alpha_{201}^{v1} \beta_2^{v2} \alpha_{210}^{v2} \alpha_{201}^{v2} \cdots \beta_2^{vK} \alpha_{210}^{vK} \alpha_{201}^{vK} \\ \beta_3^{v1} \alpha_{310}^{v1} \alpha_{301}^{v1} \beta_3^{v2} \alpha_{310}^{v2} \alpha_{301}^{n?} \cdots \beta_3^{vK} \alpha_{310}^{vK} \alpha_{301}^{vK} \\ \cdots \cdots \cdots \cdots \cdots \cdots \cdots \cdots \cdots \\ \beta_F^{v1} \alpha_{F10}^{v1} \alpha_{F01}^{v1} \beta_F^{v2} \alpha_{F10}^{v2} \alpha_{F01}^{v2} \cdots \beta_F^{vK} \alpha_{F10}^{vK} \alpha_{F01}^{vK} \end{bmatrix} \tag{2.82}$$

Collect the motion and shape parameters in the $2(F-1) \times 3$ matrix M and in the $3 \times 3K$ matrix S as follows:

$$M = \begin{bmatrix} i_{x2} i_{x3} \cdots i_{xF} j_{x2} j_{x3} \cdots j_{xF} \\ i_{y2} i_{y3} \cdots i_{yF} j_{y2} j_{y3} \cdots j_{yF} \\ i_{z2} i_{z3} \cdots i_{zF} j_{z2} j_{z3} \cdots j_{zF} \end{bmatrix}^T, \tag{2.83}$$

$$S^T = \begin{bmatrix} x_0^1 & 1 & 0 & x_0^2 & 1 & 0 & \cdots x_0^K & 1 & 0 \\ y_0^1 & 0 & 1 & y_0^2 & 0 & 1 & \cdots y_0^K & 0 & 1 \\ a_{00}^1 & a_{10}^1 & a_{01}^1 & a_{00}^2 & a_{10}^2 & a_{01}^2 & \cdots a_{00}^K & a_{10}^K & a_{01}^K \end{bmatrix}. \tag{2.84}$$

With these definitions, we write, according to the system of Equations (2.74–2.79)

$$R = MS^T. \tag{2.85}$$

The matrix of 2-D motion parameters R is $2 (F - 1) \times 3K$, but it is rank deficient. In a noiseless situation, R is rank 3 reflecting the high redundancy in the data due to the 3-D rigidity of the object. This is a restatement of the rank deficiency that has been reported in [12] for a matrix that collects image feature positions.

2.4.3.5 Rank 1 factorization

Estimating the shape and position parameters given the observation matrix R is a nonlinear LS problem. This problem has a specific structure: it is a bilinear constrained LS problem. The bilinear relation comes from (2.85) and the constraints are imposed by the orthonormality of the rows of the matrix M (2.83). We find a suboptimal solution to this problem in two stages. The first stage, decomposition stage, solves the unconstrained bilinear problem $R = MS^T$ in the LS sense. The second stage, normalization stage, computes a normalizing factor by approximating the constraints imposed by the structure of the matrix M.

2.4.3.6 Decomposition stage

Because the first two rows of S^T are known, we show that the unconstrained bilinear problem $R = MS^T$ is solved by the factorization of a rank 1 matrix $\tilde{R}$ rather than a rank 3 matrix as in [12] (see [14] for the details). Define $M = [M_0, m_3]$ and $S = [S_0, a]$. M_0 and S_0 contain the first two columns of M and S, respectively; m_3 is the third column of M, and a is the third column of S. We decompose the shape parameter vector a into the component that belongs to the space spanned by the columns of S_0 and the component orthogonal to this space as

$$a = S_0 b + a_1, \quad \text{with} \quad a_1^T S_0 = [0 \ 0]. \tag{2.86}$$

The matrix $\tilde{R}$ is R multiplied by the orthogonal projector onto the orthogonal complement of the space spanned by the first two columns of S,

$$\tilde{R} = R \left[I - S_0 \left(S_0^T S_0 \right)^{-1} S_0^T \right]. \tag{2.87}$$

This projection reduces the rank of the problem from 3 (matrix R) to 1 (matrix $\tilde{R}$). In fact (see [14]), we get

$$\tilde{R} = m_3 a_1^T. \tag{2.88}$$

The solution for m_3 and a_1 is given by the rank 1 matrix that best approximates $\tilde{R}$. In a noiseless situation, $\tilde{R}$ is rank 1. By computing the largest singular value of $\tilde{R}$ and the associated singular vectors, we get

$$\tilde{R} \doteq u\sigma v^T, \quad \hat{m}_3 = \alpha\,u, \quad \hat{a}_1^T = \frac{\sigma}{\alpha}v^T, \tag{2.89}$$

where α is a normalizing scalar different from 0. To compute u, σ, and v, we use the power method — an efficient iterative algorithm that avoids the need to compute the complete SVD (see [82]). This makes our decomposition algorithm simpler than the one in [12].

2.4.3.7   Normalization stage

We see that decomposition of the matrix $\tilde{R}$ does not determine the vector b. This is because the component of a that lives in the space spanned by the columns of S_0 does not affect the space spanned by the columns of the entire matrix S, and the decomposition stage restricts only this last space. We compute α and b by imposing the constraints that come from the structure of matrix M. We get (see [14])

$$\hat{M} = N\begin{bmatrix} I_{2\times2} & 0_{2\times1} \\ -\alpha b^T & \alpha \end{bmatrix}, \quad \text{where } N = \left[RS_0\left(S_0^T S_0\right)^{-1} u \right]. \tag{2.90}$$

The constraints imposed by the structure of M are the unit norm of each row and the orthogonality between row j and row $j + F - 1$. In terms of N, α, and b, the constraints are

$$n_i^T\begin{bmatrix} I_{2\times2} & -\alpha b \\ -\alpha b^T & \alpha^2(1 + b^T b) \end{bmatrix} n_i = 1, \quad 1 \le i \le 2(F-1), \tag{2.91}$$

$$n_j^T\begin{bmatrix} I_{2\times2} & -\alpha b \\ -\alpha b^T & \alpha^2(1 + b^T b) \end{bmatrix} n_{j+F-1} = 0, \quad 1 \le j \le F-1, \tag{2.92}$$

where n_i^T denotes the row i of matrix N. We compute α and b from the linear LS solution of the system above in a similar way to the one described in [12]. This stage is also simpler than the one in [12] because the number of unknowns is 3 (α and $b = [b_1, b_2]^T$) as opposed to the 9 entries of a generic 3×3 normalization matrix.

2.4.3.8   Texture recovery

After recovering the 3-D shape and the 3-D motion parameters, the texture of each surface patch is estimated by averaging the video frames coregistered according to the recovered 3-D structure.

From the 3-D shape parameters $\{a_{00}^k, a_{10}^k, a_{01}^k\}$ and the 3-D motion parameters $\{t_{uf}, t_{vf}, \boldsymbol{\Theta}_f\}$, we compute the parameters $\boldsymbol{\alpha}_f^k = \{\alpha_{f00}^{uk}, \alpha_{f10}^{uk}, \alpha_{f01}^{uk}, \alpha_{f00}^{vk}, a_{f10}^{vk}, a_{f01}^{vk}\}$, through Equations (2.74–2.79). Then, we compute the texture of the planar patch k by

$$T(s) = \frac{1}{F} \sum_{f=1}^{F} I_f(u_f(s)) \tag{2.93}$$

where $u_f(s)$ is the affine mapping parameterized by $\boldsymbol{\alpha}_f^k$, given by expression (2.73). The computation in Expression (2.93) is very simple, involving only an affine warping of each image.

2.4.4   Video synthesis

The goal of the video synthesis task is to generate a video sequence from the recovered 3-D motion, 3-D shape, and texture of the object. The synthesis task is much simpler than the analysis and involves only an appropriate warping of the recovered object texture.

Each frame is synthesized by projecting the object according to model (2.69). As defined above, the point s on the surface of the object projects in frame f onto $u_f(s)$ on the image plane. For planar surface patches, we have seen in the previous section that the mapping $u_f(s)$ is a simple affine motion model, see Expression (2.73) whose parameters are directly related to the 3-D shape and 3-D motion parameters through Expressions (2.74–2.79). The operations that must be carried out to synthesize the region corresponding to surface patch k at time instant f are: from the 3-D shape parameters $\{a_{00}^k, a_{10}^k, a_{01}^k\}$ and the 3-D motion parameters $\{t_{uf}, t_{vf}, \boldsymbol{\Theta}_f\}$, compute the parameters $\boldsymbol{\alpha}_f^k = \{\alpha_{f00}^{uk}, \alpha_{f10}^{uk}, \alpha_{f01}^{uk}, \alpha_{f00}^{vk}, \alpha_{f10}^{vk}, \alpha_{f01}^{vk}\}$, through Equations (2.74–2.79); then, project the texture of the patch k according to the mapping $u_f(s)$, given by Expression (2.73).

The projection of the texture is straightforward because it involves only the warping of the texture image according to the mapping $s_f(u)$, the inverse of $u_f(s)$. In fact, according to Expressions (2.69) and (2.71), we get

$$I_f(u) = P\{M(m_f)\mathcal{O}\}(u) = T(s_f(u)), \tag{2.94}$$

where the mapping $s_f(u)$ is affine. This mapping is computed from the parameter vector $\boldsymbol{\alpha}_f^k$ by inverting Expression (2.73). We get

$$s_f(u) = s(A_f^k, t_f^k; u, v) = A_f^k \begin{bmatrix} u \\ v \end{bmatrix} + t_f^k, \tag{2.95}$$

where

$$A_f^k = \begin{bmatrix} \alpha_{f10}^{uk} & \alpha_{f01}^{uk} \\ \alpha_{f10}^{vk} & \alpha_{f01}^{vk} \end{bmatrix}^{-1}, \quad t_f^k = -\begin{bmatrix} \alpha_{f10}^{uk} & \alpha_{f01}^{uk} \\ \alpha_{f10}^{vk} & \alpha_{f01}^{vk} \end{bmatrix}^{-1}\begin{bmatrix} \alpha_{f00}^{uk} \\ \alpha_{f00}^{vk} \end{bmatrix} + \begin{bmatrix} x_0^k \\ y_0^k \end{bmatrix}. \quad (2.96)$$

2.4.5 Experiment

We used a hand-held taped video sequence of 30 frames showing a box over a carpet. The image in the top of Figure 2.25 shows one frame of the box video sequence. The 3-D shape of the scene is well described in terms of four planar patches. One corresponds to the floor, and the other three correspond to the three visible faces of the box. The camera motion was approximately a rotation around the box.

We processed the box video sequence by using our method. We start by estimating the parameters describing the 2-D motion of the brightness pattern in the image plane.

The 2-D motion in the image plane is described by the affine motion model. To segment the regions corresponding to different planar patches, we used the simple method of sliding a 20×20 window across the image and detected abrupt changes in the affine motion parameters. We estimated the affine motion parameters by using the method of [81].

From the affine motion parameters estimates, we recovered the 3-D structure of the scene by using the surface-based rank 1 factorization method outlined above. For the box video sequence, the shape matrix S contains four submatrices, one for each planar patch. Each submatrix is a matrix that contains the 3-D shape parameters of the corresponding surface patch. According to the general structure of matrix S, as specified in Subsection 2.4.3, S is the 3×12 matrix

$$S^T = \begin{bmatrix} x_0^1 & 1 & 0 & x_0^2 & 1 & 0 & x_0^3 & 1 & 0 & x_0^4 & 1 & 0 \\ y_0^1 & 0 & 1 & y_0^2 & 0 & 1 & y_0^3 & 0 & 1 & y_0^4 & 0 & 1 \\ a_{00}^1 & a_{10}^1 & a_{01}^1 & a_{00}^2 & a_{10}^2 & a_{01}^2 & a_{00}^3 & a_{10}^3 & a_{01}^3 & a_{00}^4 & a_{10}^4 & a_{01}^4 \end{bmatrix}, \quad (2.97)$$

where (x_0^k, y_0^k) are the coordinates of the centroid of the support region of patch k and $\{a_{00}^k, a_{10}^k, a_{01}^k\}$ are the parameters describing the 3-D shape of the patch by $z = a_{00}^k + a_{10}^k(x - x_0^k) + a_{01}^k(y - y_0^k)$.

We computed the parameters describing the 3-D structure, i.e., the 3-D motion parameters $\{t_{uf}, t_{vf}, \Theta_f, 1 \le f \le 30\}$ and the 3-D shape parameters $\{a_{00}^n, a_{01}^n, 1 \le n \le 4\}$ from the image motion parameters in $\{\alpha_{f10}^{uk}, \alpha_{f01}^{uk}, \alpha_{f10}^{vk}, \alpha_{f01}^{vk}, \alpha_{f00}^{uk}\alpha_{f00}^{vk}, 1 \le f \le 30, 1 \le k \le 4\}$, by using the surface-based rank 1 factorization method. After computing the 3-D structure parameters, we recover the texture of each surface patch by averaging the video frames coregistered according to the recovered 3-D structure, as described in Subsection 2.4.3.

Figure 2.23 shows a perspective view of the reconstructed 3-D shape with the scene texture mapped on it. The spatial limits of the planar patches were determined in the following way. Each edge that links two visible patches was computed from the intersection of the planes corresponding to the patches. Each edge that is not in the intersection of two visible patches was computed by fitting a line to the boundary that separates two regions with different 2-D motion parameters. We see that the angles between the planar patches are correctly recovered.

2.4.6   Applications

In this section, we highlight some of the potential applications of the proposed 3-D model-based video representation.

2.4.6.1   Video coding

Model-based video representations enable very low bit rate compression. Basically, instead of representing a video sequence in terms of frames and pixels, 3-D model-based approaches use the recovered 3-D structure. A video sequence is then represented by the 3-D shape and texture of the object and its 3-D motion. Within the surface-based representation, the 3-D motion and 3-D shape are coded with a few parameters and the texture is coded as a set of ordinary images, one for each planar patch. We use the box video sequence to illustrate this video compression scheme.

The video analysis task consists in recovering the object shape, object motion, and object texture from the given video. The steps of the analysis task for the box video sequence were detailed above. Figure 2.24 shows frontal views of the four elemental texture constructs of the surface-based

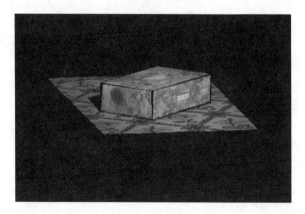

Figure 2.23 A persective view of the 3-D shape and texture reconstructed from the box video sequence.

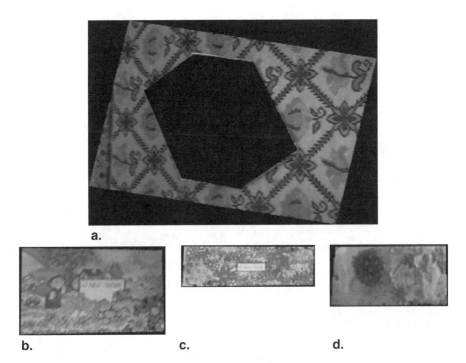

Figure 2.24 The four planar patches that consitute the elemental texture constructs: (a) the carpet (floor level); (b)–(d) from the left to the right, the three visible faces of the box: top of the box, the right side of the box, and the left side of the box.

representation of the box video sequence. On the top, the planar patch corresponding to the carpet is not complete. This is because the region of the carpet that is occluded by the box cannot be recovered from the video sequence. The other three images on the bottom of Figure 2.24 are the three faces of the box.

The video synthesis task consists in generating a video sequence from the recovered 3-D motion, 3-D shape, and texture of the object. The synthesis task is much simpler than the analysis because it involves only an appropriate warping of the recovered object texture. Each frame is synthesized by projecting the object texture as described in Subsection 2.4.4.

The original sequence has $50 \times 320 \times 240 = 3{,}840{,}000$ bytes. The representation based on the 3-D model needs

$$\sum_n T_n + \sum_n S_n + 50 \times M = \sum_n T_n + 2248$$

bytes, where T_n is the storage size of the texture of patch n, S_n is the storage size of the shape of patch n, and M is the storage size of each camera position. Since the temporal redundancy was eliminated, the compression ratio chosen for the spatial conversion governs the overall video compression ratio. To compress the texture of each surface patch in Figure 2.24, we used the JPEG standard with two different compression ratios. The storage sizes T_1, T_2, T_3, and T_4 of the texture patches were, in bytes, from left to right in Figure 2.5, 2606, 655, 662, and 502 for the higher compression ratio and 6178, 1407, 1406, and 865 for the lower compression ratio. These storage sizes lead to the average spatial compression ratios of 31:1 and 14:1.

The first frame of the original box video sequence is shown in Figure 2.25(a). Figures 2.25(b) and (c) show the first frame of the synthesized sequence for the two different JPEG spatial compression ratios. Figure 2.25(c) shows the first frame obtained with the higher spatial compression ratio, leading to the overall video compression ratio of 575:1. Figure 2.25(b) corresponds to the lower spatial compression ratio and an overall video compression ratio of 317:1. In both cases, the compression ratio due to the elimination of the temporal redundancy is approximately 20.

From the compressed frames in Figure 2.25, we see that the overall quality is good but there are small artifacts in the boundaries of the surface patches.

2.4.6.2   Video content addressing

Content-based addressing is an important application of the 3-D model-based video representation. Current systems that provide con-

a.

b.   c.

Figure 2.25 Video compression. (a) frame 1 of the box video sequence: (b) frame 1 of the synthesized sequence for a compression ratio of 575:1; (c) frame 1 of the synthesized sequence coded for a compression ratio of 317:1.

tent-based access work by first segmenting the video in a sequence of shots and then labeling each shot with a distinctive indexing feature. The most common features used are image-based features, such as color histograms or image moments. By using 3-D models, we improve both the temporal segmentation and the indexing. The temporal segmentation can account for the 3-D content of the scene. Indexing by 3-D features directly related to the 3-D shape enables queries by object similarity. See [11] for illustrative examples of the use of 3-D models in digital video processing.

2.4.6.3   *Virtualized reality*

Current methods to generate virtual scenarios are expensive. Either 3-D models are generated in a manual way, which requires a human to specify the details of the 3-D shape and texture, or auxiliary equipment, like a laser range finder, needs to be used to capture the 3-D reality. Our work can be used to generate automatically virtual reality scenes. The 3-D models obtained from the real life video data can be used to build synthetic image sequences. The synthesis is achieved by specifying the sequence of viewing positions along time. The viewing positions are arbitrary; they are specified by the user, either in an interactive way or by an automatic procedure. For example, the images in Figure 2.26 were obtained by rotating the 3-D model represented in Figure 2.23. Note that only the portion of the model that was seen in the original video sequence is synthesized in the views of Figure 2.26. Other views are

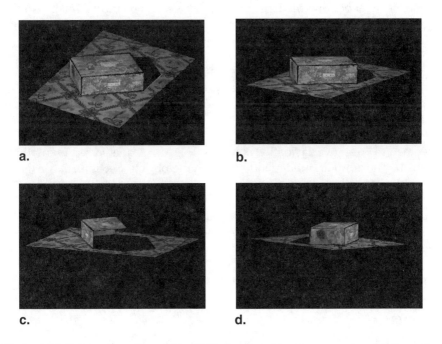

a.   b.

c.   d.

Figure 2.26 Perspective views of the 3-D shape and texture reconstructed from the box video sequence of Figure 2.2.

generated in a similar way. Synthetic images are obtained by selecting from these views a rectangular window, corresponding to the camera field of view. This is an example of virtual manipulation of real objects. More complex scenes are obtained by merging real objects with virtual entities.

2.4.7   Summary

In this section we described a framework for 3-D content-based digital video representation. Our framework represents a video sequence in terms of the 3-D rigid shape of the scene (a piecewise planar surface), the texture of the scene, and the 3-D motion of the camera. When analyzing a video sequence, we recover 3-D rigid models by using a robust factorization approach. When synthesizing a video sequence, we simply warp the surface patches that describe the scene, avoiding the use of computationally expensive rendering tools. One experiment illustrates the performance of the algorithms used. We highlight potential applications of the proposed 3-D model-based video representation framework.

2.5   Conclusion

In this chapter, we presented 2-D and 3-D content-based approaches to the representation of video sequences. These representations are quite efficient; they reduce significantly the amount of data needed to describe the video sequence by exploiting the large overlap between consecutive images in the video sequence. These representations abstract from the video sequence the informational units — constructs — from which the original video can be regenerated. Content based video representations can be used in compression, with the potential to achieve very large compression ratios [61, 62]. But their applications go well beyond compression, from video editing (see [63]) to efficient video database indexing and querying. The approaches presented in this chapter generalize the traditional mosaics, going beyond panoramic views of the background. In GV, presented in Section 2.3, the world is assumed to be an overlay of 2-D objects. GV leads to layered representations of the background and independently moving objects. Section 2.3 considers also 3-D layered mosaics. Section 2.4 describes the surface-based rank 1 factorization method that constructs 3-D representations of objects from monocular video sequences. An essential preprocessing step in obtaining these video representations is the segmentation of the objects of interest from the background. Section 2.2 overviews several energy minimization-type approaches to the problem of image segmentation — an essential step in constructing these content-based video representations.

References

[1]   Lippman, A.: Movie maps: an application of the optical videodisc to computer graphics. Proceedings of SIGGRAPH, ACM (1980) 32–43

[2]   Burt, P.J., Adelson, E.H.: A multiresolution spline with application to image mosaics. ACM Transactions on Graphics **2** (1983) 217–236

[3]   Hansen, M., Anandan, P., Dana, K., van der Wal, G., Burt, P.J.: Real-time scene stabilization and mosaic construction. Proceedings of the ARPA APR Workshop (1994) 41–49

[4]   Teodosio, L., Bender, W.: Salient video stills: Content and context preserved. Proceedings of the First ACM International Conference on Multimedia (1993) 39–46

[5]   Irani, M., Anandan, P., Bergen, J., Kumar, R., Hsu, S.: Efficient representations of video sequences and their applications. Signal Processing: Image Communication **8** (1996) 327–351

[6]   Wang, J., Adelson, E.: Representing moving images with layers. IEEE Transactions Image Processing **3** (1994) 625–638

[7]   Kumar, R., Anandan, P., Irani, M., Bergen, J.R., Hanna, K.J.: Representation of scenes from collections of images. IEEE Workshop on Representations of Visual Scenes (1995) 10–17

[8]   Aizawa, K., Harashima, H., Saito, T.: Model-based analysis-synthesis image coding (MBASIC) system for a person's face. Signal Processing: Image Communication **1** (1989) 139–152

[9]   Kaneko, M., Koike, A., Hatori, Y.: Coding of a facial image sequence based on a 3-D model of the head and motion detection. Journal of Visual Communication and Image Representation **2** (1991) 39–54

[10] Soucy, M., Laurendeau, D.: A general surface approach to the integration of a set of range views. IEEE Transactions on Pattern Analysis and Machine Intelligence **17** (1995) 344–358

[11] Martins, F.C.M., Moura, J.M.F.: Video representation with three-dimensional entities. IEEE Journal on Selected Areas in Communications **16** (1998) 71–85

[12] Tomasi, C., Kanade, T.: Shape and motion from image streams under orthography: a factorization method. International Journal of Computer Vision **9** (1992) 137–154

[13] Aguiar, P.M.Q., Moura, J.M.F.: Three-dimensional modeling from two-dimensional video. IEEE Transactions on Image Processing **10** (2001) 1541–1551

[14] Aguiar, P.M.Q., Moura, J.M.F.: Rank 1 weighted factorization for 3 D structure recovery: algorithms and performance analysis. IEEE Transactions on Pattern Analysis and Machine Intelligence **25** (2003) 1134–1149

[15] Kanade, T., Rander, P., Narayanan, J.: Virtualized reality: constructing virtual worlds from real scenes. IEEE Multimedia **4** (1997) 34–47

[16] Debevec, P., Taylor, C., Malik, J.: Modeling and rendering architecture from photographs: a hybrid geometry- and image-based approach. SIGGRAPH International Conference on Computer Graphics and Interactive Techniques (1996) 11–20

[17] Adelson, E., Bergen, J.: The pleenoptic function and the elements of early vision. In Movshon, J., ed.: Computational Models of Visual Processing. Cambridge, MA: MIT Press (1991)

[18] Levoy, M., Hanrahan, P.: Light field rendering. SIGGRAPH International Conference on Computer Graphics and Interactive Techniques (1996) 31–42

[19] Gortler, S., Grzeszczuk, R., Szeliski, R., Cohen, M.: The lumigraph. SIGGRAPH International Conference on Computer Graphics and Interactive Techniques (1996) 43–54

[20] Kass, M., Witkin, A., Terzopoulos, D.: Snakes: active contour models. International Journal of Computer Vision **1** (1988) 321–331

[21] Elsgolc, L.E.: Calculus of Variations. Reading, MA: Addison-Wesley Publishing Company (1962)

[22] Castleman, K.R.: Digital Image Processing. Upper Saddle River, NJ: Prentice Hall (1996)

[23] Gonzalez, R.C., Woods, R.E.: Digital Image Processing. Reading, MA: Addison-Wesley Publishing Company (1993)

[24] Sobel, I.: Neighborhood coding of binary images for fast contour following and general array binary processing. Computer Graphics and Image Processing **8** (1978) 127–135

[25] Prewitt, J.M.S.: Object enhancement and extraction. In Lipkin, B.S., Rosenfeld, A., eds.: Picture Processing and Psychopictories. New York: Academic Press (1970)

[26] Marr, D., Hildreth, E.: Theory of edge detection. Proceedings Royal Society of London **B 207** (1980) 187–217

[27] Torre, V., Poggio, T.A.: On edge detection. IEEE Transactions on Pattern Analysis and Machine Intelligence **8** (1986) 147–163

[28] Canny, J.F.: A computational approach to edge detection. IEEE Transactions on Pattern Analysis and Machine Intelligence **8** (1986) 769–798

[29] Casadei, S., Mitter, S.: Hierarchical image segmentation — part I: detection of regular curves in a vector graph. International Journal of Computer Vision **27** (1988) 71–100

[30] Casadei, S., Mitter, S.: Beyond the uniqueness assumption: ambiguity representation and redundancy elimination in the computation of a covering sample of salient contour cycles. Computer Vision and Image Understanding **76** (1999) 19–35

[31] Adams, R., Bischof, L.: Seeded region growing. IEEE Transactions on Pattern Recognition and Machine Intelligence **16** (1994) 641–647

[32] Leonardis, A., Gupta, A., Bajcsy, R.: Segmentation of range images as the search for geometric parametric models. International Journal of Computer Vision **14** (1995) 253–277

[33] Pavlidid, T., Liow, Y.T.: Integrating region growing and edge detection. IEEE Transactions on Pattern Analysis and Machine Intelligence **12** (1990) 225–233

[34] Terzopoulos, D., Witkin, A.: Constraints on deformable models: recovering shape and non-rigid motion. Artificial Intelligence **36** (1988) 91–123

[35] Cohen, L.D.: On active contour models and balloons. CVGIP: Image Understanding **53** (1991) 211–218

[36] Ronfard, R.: Region-based strategies for active contour models. International Journals on Computer Vision **13** (1994) 229–251

[37] Malladi, R., Sethian, J.A., Vemuri, B.: Shape modeling with front propagation: a level set approach. IEEE Transactions on Pattern Analysis and Machine Intelligence **17** (1995) 158–175

[38] Xu, C., Prince, J.L.: Snakes, shapes, and gradient vector flow. IEEE Transactions on Medical Imaging **7** (1998) 359–369

[39] Chakraborty, A., Staib, L., Duncan, J.: Deformable boundary finding in medical images by integrating gradient and region information. IEEE Transactions on Medical Imaging **15** (1996) 859–570

[40] Yezzi, A., Kichenassamy, S., Kumar, A., Olver, P., Tannenbaum, A.: A geometric snake model for segmentation of medical imagery. IEEE Transactions on Medical Imaging **16** (1997) 199–209

[41] Chakraborty, A., Duncan, J.: Game-theoretic integration for image segmentation. IEEE Transactions on Pattern Analysis and Machine Intelligence **21** (1999) 12–30

[42] Chan, T.F., Vese, L.A.: Active contours without edges. IEEE Transactions on Image Processing **10** (2001) 266–277

[43] Morel, J.M., Solimini, S.: Variational Methods in Image Segmentation. Cambridge, MA: Birkhauser (1995)

[44] Zabusky, N.J., Overman II, E.A.: Tangential regularization of contour dynamical algorithms. Journal of Computational Physics **52** (1983) 351–374

[45] Osher, S., Sethian, J.A.: Fronts propagating with curvature-dependent speed: algorithms based on Hamilton–Jacobi formulations. Journal of Computational Physics **79** (1988) 12–49

[46] Sethian, J.A.: Level Set Methods and Fast Marching Methods. Cambridge, U.K.: Cambridge University Press (1999)

[47] Caselles, V., Kimmel, R., Sapiro, G.: Geodesic active contours. Proceedings of the 5th International Conference on Computer Vision (ICCV-95) (1995) 694–699

[48] Mumford, D., Shah, J.: Optimal approximations by piecewise smooth functions and associated variational problems. Communications in Pure and Applied Mathematics **12** (1989)

[49] Xu, C., Prince, J.L.: Gradient vector flow: a new external force for snakes. IEEE Proceedings Conference on Computer Vision and Pattern Recognition (CVPR) (1997) 66–71

[50] McInerney, T., Terzopoulos, D.: Topologically adaptable snakes. Proceedings of the International Conference on Computer Vision (1995) 840–845

[51] Sapiro, G.: Geometric Partial Differential Equations and Image Analysis. Cambridge, U.K.: Cambridge University Press (2001)

[52] Osher, S.: Riemann solvers, the entropy condition, and difference approximations. SIAM Journal of Numerical Analysis **21** (1984) 217–235

[53] Sethian, J.A.: Numerical algorithms for propagating interfaces: Hamilton-Jacobi equations and conservation laws. Journal of Differential Geometry **31** (1990) 131–161

[54] Caselles, V., Kimmel, R., Sapiro, G.: Geodesic snakes. International Journal of Computer Vision **22** (1997) 61–79

[55] Sethian, J.A.: Curvature and evolutions of fronts. Communications in Mathematics and Physics **101** (1985) 487–499

[56] McInerney, T., Terzopoulos, D.: T-snake: topology adaptive snakes. Medical Image Analysis **4** (2000) 73–91

[57] Borgefors, G.: Distance transformations in arbitrary dimensions. Computer Vision, Graphics, and Image Processing **27** (1984) 321–345

[58] Alvarez, L., Guichard, F., Lions, P.L., Morel, J.M.: Axioms and fundamental equations of image processing. Archive for Rational Mechanics **123** (1993) 199–257

[59] Pluempitiwiriyawej, C., Moura, J.M.F., Wu, Y.J.L., Kanno, S., Ho, C.: Stochastic active contour for cardiac MR image segmentation. IEEE International Conference on Image Processing (ICIP) (2003) 1097–1100

[60] Pluempitiwiriyawej, C., Moura, J.M.F., Wu, Y.J.L., Kanno, S., Ho, C.: STACS: New active contour scheme for cardiac MR image segmentation. Submitted for publication. 25 pages (2003)

[61] Jasinschi, R.S., Moura, J.F.M., Cheng, J.C., Asif, A.: Video compression via constructs. Proceedings of IEEE ICASSP **4** (1995) 2165–2168

[62]   Jasinschi, R.S., Moura, J.F.M.: Content-based video sequence representation. Proceedings of IEEE ICIP **2** (1995) 229–232

[63]   Jasinschi, R.S., Moura, J.F.M.: Nonlinear video editing by generative video. Proceedings of IEEE ICASSP **II** (1996) 1220–1223

[64]   Jasinschi, R.S., Moura, J.M.F.: Generative video: very low bit rate video compression U.S. Patent 5,854,856 (1998)

[65]   Jasinschi, R.S., Naveen, T., Babic-Vovk, P., Tabatabai, A.: Apparent 3-D camera velocity extraction and its application. PCS'99 (1999)

[66]   Jasinschi, R.S., Naveen, T., Babic-Vovk, P., Tabatabai, A.: Apparent 3-D camera velocity: extraction and applications. IEEE Transactions on Circuits and Systems for Video Technology **10** (2000) 1185–1191

[67]   Jasinschi, R.S., Tabatabai, A., Thumpudi, N., Babic-Vov, P.: 2-D extended image generation from 3-D data extracted from a video sequence U.S. Patent 6,504,569. (2003)

[68]   Longuett-Higgins, H.C.: A computer algorithm for reconstructing a scene from two projections. Nature **293** (1981) 133–135

[69]   Tsai, R.Y., Huang, T.S.: Uniqueness and estimation of three-dimensional motion parameters of rigid objects with curved surfaces. Transactions on Pattern Analysis and Machine Intelligence **6** (1984) 13–27

[70]   Hartley, R.I.: In defense of the eight-point algorithm. IEEE Transactions on Pattern Recognition and Machine Intelligence **19** (1997) 580–593

[71]   Aguiar, P.M.Q., Moura, J.M.F.: Fast 3-D modelling from video. IEEE Workshop on Multimedia Signal Processing, Copenhagen, Denmark (1999)

[72]   Broida, T., Chellappa, R.: Estimating the kinematics and structure of a rigid object from a sequence of monocular images. IEEE Transactions on Pattern Analysis and Machine Intelligence **13** (1991) 497–513

[73]   Azarbayejani, A., Pentland, A.P.: Recursive estimation of motion, structure, and focal length. IEEE Transactions on Pattern Analysis and Machine Intelligence **17** (1995) 562–575

[74]   Poelman, C.J., Kanade, T.: A paraperspective factorization method for shape and motion recovery. IEEE Transactions on Pattern Analysis and Machine Intelligence **19** (1997) 206–218

[75]   Quan, L., Kanade, T.: A factorization method for affine structure from line correspondences. IEEE International Conference on Computer Vision and Pattern Recognition, San Francisco, CA (1996) 803–808

[76]   Morita, T., Kanade, T.: A sequential factorization method for recovering shape and motion from image streams. IEEE Transactions on Pattern Analysis and Machine Intelligence **19** (1997) 858–867

[77]   Costeira, J.P., Kanade, T.: A factorization method for independently moving objects. International Journal of Computer Vision **29** (1998) 159–179

[78]   Ayache, N.: Artificial Vision for Mobile Robots. Cambridge, MA: MIT Press (1991)

[79]   Zheng, H., Blostein, S.D.: Motion-based object segmentation and estimation using the MDL principle. IEEE Transactions on Image Processing **4** (1995) 1223–1235

[80]   Chang, M., Tekalp, M., Sezan, M.: Simultaneous motion estimation and segmentation. IEEE Transactions on Image Processing **6** (1997) 1326–1333

[81]   Bergen, J.R., Anandan, P., Hanna, K.J., Hingorani, R.: Hierarchical model-based motion estimation. European Conference on Computer Vision, Italy (1992) 237–252

[82]   Golub, G.H., Van-Loan, C.F.: Matrix Computations. Baltimore: Johns Hopkins University Press (1996)

chapter 3

The computation of motion

Christoph Stiller, Sören Kammel, Jan Horn, and Thao Dang

Contents

Abstract. Motion computation is a key element to numerous applications. Image coding, velocity sensing, image alignment, and object detection are but a few examples. This chapter focuses on methods for the computation of motion. The formation process of motion in image sequences induced from 3-D motion in the real world is analyzed. Several constraints apply to 2-D motion of arbitrarily shaped, but rigid, 3-D objects. We discuss different approaches to motion computation and outline their underlying model assumptions. These models describe the interdependence of motion and image intensity variation, the spatio-temporal variation of motion vectors, and the shape of coherently moving regions. Uncertainty assessment plays an important role in any

0-8493-1526-3/2004/$0.00+$1.50

measurement. Covariance provides a simple yet efficient means to describe uncertainty and should accompany each motion measurement. We show how it can be integrated naturally into the motion computation process and propagated between the processing steps. A focus of this chapter is on practical considerations for motion computation. We discuss four concise applications from different fields. Beyond the large differences in requirements, motion computation can be formulated in a unique mathematical framework. The individual applications take benefit of uncertainty information in different ways. In image sequence coding uncertainty quantifies the content of information that could be gathered and is used to prioritize the information to be transmitted. In 3-D sensing uncertainty is employed for fusion of individual measurements and tracking of object parameters.

3.1   The importance of motion information in image processing

Motion information plays a key role in numerous image processing tasks. 2-D motion in imagery is induced by 3-D motion of projected objects or movements of the camera. Roughly spoken, from a signal-processing point of view, motion trajectories represent the spatio-temporal directions of the strongest intensity resemblance in image sequences. From a measurement point of view, 2-D motion in imagery is an observation incorporating information about 3-D geometry and 3-D motion of the real world.

Applications that include estimation of motion are manifold. The following examples shall illustrate their broad spectrum without the aim of completeness. In image sequence encoding, motion information serves to remove spatio-temporal redundancy from image data. Motion compensated prediction reconstructs the major temporal variations in imagery based on a few transmitted motion parameters. Digital filters, e.g., for noise reduction, interpolation, or restoration, are oriented in the direction of motion trajectories for highest efficiency [1]. In scene analysis, the correspondences of image positions over time that are established by 2-D motion information serve the measurement of quantities like 3-D shape, 3-D position, and 3-D motion of objects in the real world. In mosaicing and image registration, motion information is extracted from image sequences to produce images with enlarged fields of view or enhanced spatial resolution.

Higher animals and humans possess the ability of seemingly effortless and instantaneous extraction of motion information. In contrast, artificial vision systems typically devote a dominant portion of computational resources to motion estimation and accept substantial computation time. Despite the huge variety of algorithms proposed in literature, motion estimation is still the subject of ongoing research. Methods of choice strongly depend on the envisaged application and accompanying processing modules.

This paper discusses motion computation from both a signal-processing and a measurement point of view. We are convinced that both worlds may be highly inspiring to each other. The remainder of this chapter is organized as follows. The next section outlines motion formation and introduces a mathematical problem description. Section 3.3 discusses important classes of motion estimators and their underlying models. A prominent issue in this context is to define an appropriate estimation criterion. Motion uncertainty measures are introduced as valuable information accompanying each motion estimate. The transfer of the analysis to four concise applications ranging from image sequence coding to production assessment is presented in Section 3.4 before the paper closes with conclusions and a discussion of prospective future developments.

3.2 Motion formation

In this paper, we focus on the most prominent source for the formation of visible motion in image sequences, namely the 3-D motion of a point in the real world that is projected onto the image plane. Let us consider a point moving over time with respect to some 3-D coordinate system. Then, its 3-D position as a function over time t describes a 3-D trajectory $\mathbf{X}(t) = (X(t), Y(t), Z(t))^{\mathsf{T}}$. From this trajectory, 3-D motion between two time instances t, $t+1$ may be readily defined as

$$\mathbf{D}(t) = \mathbf{X}(t+1) - \mathbf{X}(t). \tag{3.1}$$

Let us now consider an image acquisition system that projects the scene onto its image plane $\mathbf{x} = (x, y)^{\mathsf{T}}$ as

$$\mathbf{x}(t) - \pi\big(\mathbf{X}(t)\big), \tag{3.2}$$

where $\pi(\cdot)$ denotes projection. It is worth noting that one loses at least one dimension of the spatial scene information through this projection of the 3-D world onto the 2-D image plane and many of the challenges in image processing arise from this loss. Recovery of the lost information, to some extent, states an *inverse* and often an *ill-posed* problem (see, e.g., [2]).

When we assume that the point under consideration does not experience occlusion at either of the time instances, the 3-D motion vector is implicitly projected onto the image plane as a 2-D motion vector as depicted in Figure 3.1

$$\mathbf{d}(t) = \pi\big(\mathbf{X}(t+1)\big) - \pi\big(\mathbf{X}(t)\big). \tag{3.3}$$

In practice, one is interested not only in the motion of a single point but also in the motion of each point that is visible in the first image. Thus, one is left with a motion vector field $\mathbf{d}(\mathbf{x}, t)$. This field is well defined for all spatio-temporal image positions $\mathbf{x} = \pi(\mathbf{X}(t))$ projecting 3-D points $\mathbf{X}$ that do

Figure 3.1 3-D motion in the real world is projected as 2-D motion onto the image plane.

not experience occlusion at time t+1. The remaining set of image positions with undefined 2-D motion is referred to as *occlusion area*.

In the following, we confine ourselves to the particular projection of a pinhole camera known as perspective projection, which is a suitable model for most cameras. We consider the coordinate system of a calibrated camera; i.e., in the mathematical framework of projective geometry, we can express perspective projection as

$$\lambda\begin{pmatrix} \mathbf{x}(t) \\ 1 \end{pmatrix} = \mathbf{X}(t), \tag{3.4}$$

where λ denotes an appropriate real scale factor. Obviously, this description of perspective projection can be converted to the form of Equation (3.2) by resolving $\lambda = Z$:

$$\mathbf{x}(t) = \pi\big(\mathbf{X}(t)\big) = \frac{1}{Z(t)}\begin{pmatrix} X(t) \\ Y(t) \end{pmatrix}. \tag{3.5}$$

Figure 3.2 depicts a point of a rigid object in the world that is viewed by a camera at time instances t and t+1. During this time interval the object has moved with respect to the camera, or, equivalently, the camera has moved with respect to the object as depicted in the figure. For the sake of brevity we introduce subscripts 1 and 2 for all positions corresponding to time t and t+1, respectively, e.g., $\mathbf{x}_1 = \mathbf{x}(t)$. The 3-D position of any point of the rigid object in camera coordinates for the two time instances is related as

$$\mathbf{X}_2 = \mathbf{R}\mathbf{X}_1 + \mathbf{t}, \tag{3.6}$$

where $\mathbf{R}$ is an orthonormal[1] 3×3 rotation matrix and $\mathbf{t}$ is a 3×1 vector of translatory movement. Insertion of the projection, Equation (3.4), yields

[1] I.e. $\mathbf{R}^{-1} = \mathbf{R}^{\mathrm{T}}$

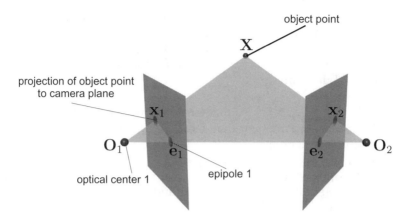

Figure 3.2 The epipolar constraint for a point **X** in the real world viewed by two cameras: The coordinates of $\mathbf{x}_2$ are restricted to the epipolar plane defined by the camera positions at times t, $t + 1$, and the image point $\mathbf{x}_1$.

$$\lambda_2 \begin{pmatrix} \mathbf{x}_2 \\ 1 \end{pmatrix} = \lambda_1 \mathbf{R} \begin{pmatrix} \mathbf{x}_1 \\ 1 \end{pmatrix} + \mathbf{t}, \tag{3.7}$$

with appropriate real scale factors λ_1 and λ_2. Hence, $\left(\mathbf{x}_2^{\mathsf{T}}, 1\right)^{\mathsf{T}}$ is a linear combination of $\mathbf{R}\left(\mathbf{x}_1^{\mathsf{T}}, 1\right)^{\mathsf{T}}$ and $\mathbf{t}$, i.e., the vector $\mathbf{v} = \mathbf{t} \times \mathbf{R}\left(\mathbf{x}_1^{\mathsf{T}}, 1\right)^{\mathsf{T}}$ is orthogonal to $\left(\mathbf{x}_2^{\mathsf{T}}, 1\right)^{\mathsf{T}}$

$$\begin{pmatrix} \mathbf{x}_2 \\ 1 \end{pmatrix}^{\mathsf{T}} \left(\mathbf{t} \times \mathbf{R} \begin{pmatrix} \mathbf{x}_1 \\ 1 \end{pmatrix} \right) = 0. \tag{3.8}$$

This finding is illustrated in Figure 3.2.

Introducing the matrix that satisfies $\mathbf{TX} = \mathbf{t} \times \mathbf{X}$, for any vector $\mathbf{X}$, finally yields the *epipolar constraint* (see, e.g., [9])

$$\mathbf{T} = [\mathbf{t}]_\times = \begin{pmatrix} 0 & -t_Z & t_Y \\ t_Z & 0 & -t_X \\ -t_Y & t_X & 0 \end{pmatrix}; \text{ with } \mathbf{t} = \begin{pmatrix} t_X \\ t_Y \\ t_Z \end{pmatrix}, \tag{3.9}$$

$$\begin{pmatrix} \mathbf{x}_2 \\ 1 \end{pmatrix}^{\mathsf{T}} \mathbf{E} \begin{pmatrix} \mathbf{x}_1 \\ 1 \end{pmatrix} = 0, \tag{3.10}$$

where $\mathbf{E} = \mathbf{TR}$ denotes the *essential matrix*.

The epipolar constraint in Equation (3.10) restricts all 2-D motion vectors of a rigid object to a single dimension. The essential matrix $\mathbf{E}$ incorporates the 3-D motion of the object with respect to the camera. It determines fans of *epipolar lines* in the first and second image, respectively. Corresponding points $\mathbf{x}_1$, $\mathbf{x}_2$ must lie on a corresponding line pair that satisfies Equation (3.10).

Rigid motion at least dominates nonrigid motion for most objects. Therefore, motion decomposition of each object into three parts is useful for motion estimation and description:

- 3-D rigid motion, i.e., $\mathbf{R}$, $\mathbf{t}$, determining the essential matrix.
- 1-D motion component that satisfies the epipolar constraint (3.10). This component specifies the 3-D shape of rigid objects and hence will likely show consistency over time.
- 1-D motion component perpendicular to the epipolar line. This component describes nonrigid motion and can therefore be expected to be small for most objects.

Further constraints are introduced if we consider three camera images (see Figure 3.3) or, equivalently, three consecutive views of a rigid object by a camera. Each pair of cameras defines an individual epipolar constraint, namely

$$\begin{pmatrix}\mathbf{x}_1\\1\end{pmatrix}^{\mathrm{T}}\mathbf{E}_{12}\begin{pmatrix}\mathbf{x}_2\\1\end{pmatrix}=0,\quad\begin{pmatrix}\mathbf{x}_2\\1\end{pmatrix}^{\mathrm{T}}\mathbf{E}_{23}\begin{pmatrix}\mathbf{x}_3\\1\end{pmatrix}=0,\quad\begin{pmatrix}\mathbf{x}_3\\1\end{pmatrix}^{\mathrm{T}}\mathbf{E}_{31}\begin{pmatrix}\mathbf{x}_1\\1\end{pmatrix}=0,\qquad(3.11)$$

where $\mathbf{E}_{ij}$ represents the essential matrix between cameras i and j. These three constraints are not independent, as is intuitively clear from Figure 3.3. Generally, given the 3-D transformations between frames $1\leftrightarrow2$ and $2\leftrightarrow3$, we can also compute the geometry between cameras 3 and 1. This geometric relation between three images can be encapsulated elegantly in the *trifocal*

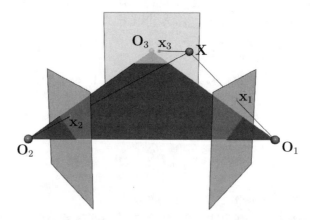

Figure 3.3 Trifocal geometry: Here, we have three individual epipolar constraints — among cameras $O_1\leftrightarrow O_2$, $O_2\leftrightarrow O_3$, and $O_3\leftrightarrow O_1$.

tensor **T**. The trifocal tensor was introduced in [3]. It is represented as a set of three 3×3 matrices

$$\mathbf{T}_k = \mathbf{t}_2 \mathbf{R}_3^{i^T} - \mathbf{R}_2^i \mathbf{t}_3^T, \quad k = 1\ldots3, \tag{3.12}$$

where $(\mathbf{R}_2, \mathbf{t}_2)$, $(\mathbf{R}_3, \mathbf{t}_3)$ denote the transformation from the coordinate system of camera 1 into the coordinate systems 2 and 3, respectively, and $\mathbf{R}_3^i$ represents the ith column of the rotation matrix $\mathbf{R}_3$.

Using the trifocal tensor, an algebraic constraint can be formulated that must be met by all triplets $(\mathbf{x}_1, \mathbf{x}_2, \mathbf{x}_3)$ of corresponding points in three images. This constraint is given by

$$\begin{bmatrix} x_2 \\ y_2 \\ 1 \end{bmatrix}_\times \left(x_1 \mathbf{T}_1 + y_1 \mathbf{T}_2 + \mathbf{T}_3 \right) \begin{bmatrix} x_3 \\ y_3 \\ 1 \end{bmatrix}_\times = 0_{3\times3}. \tag{3.13}$$

It can be shown that only four of the nine relations in Equation (3.13) are linearly independent (see [4]). An important application of the trifocal tensor is the *transfer problem*. Given the positions $\mathbf{x}_1$ and $\mathbf{x}_2$ of a point in two camera frames, predict the coordinates $\mathbf{x}_3$ of the point in a third image. If we denote each point in so-called projective coordinates $\tilde{\mathbf{x}}_k = \lambda \cdot [\tilde{x}_{k1}, \tilde{x}_{k2}, \tilde{x}_{k3}]^T = [x_k, y_k, 1]^T$ with some real λ, it can be shown that

$$x_3 = \frac{\tilde{x}_{2i}\left(\tilde{x}_{11}T_{1j1} + \tilde{x}_{12}T_{2j1} + T_{3j1}\right) - \tilde{x}_{2j}\left(\tilde{x}_{11}T_{1i1} + \tilde{x}_{12}T_{2i1} + T_{3i1}\right)}{\tilde{x}_{2i}\left(\tilde{x}_{11}T_{1j3} + \tilde{x}_{12}T_{2j3} + T_{3j3}\right) - \tilde{x}_{2j}\left(\tilde{x}_{11}T_{1i3} + \tilde{x}_{12}T_{2i3} + T_{3i3}\right)},$$

$$\tag{3.14}$$

$$y_3 = \frac{\tilde{x}_{2i}\left(\tilde{x}_{11}T_{1j2} + \tilde{x}_{12}T_{2j2} + T_{3j2}\right) - \tilde{x}_{2j}\left(\tilde{x}_{11}T_{1i2} + \tilde{x}_{12}T_{2i2} + T_{3i2}\right)}{\tilde{x}_{2i}\left(\tilde{x}_{11}T_{1j3} + \tilde{x}_{12}T_{2j3}' + T_{3j3}\right) - \tilde{x}_{2j}\left(\tilde{x}_{11}T_{1i3} + \tilde{x}_{12}T_{2i3} + T_{3i3}\right)},$$

for all $i, j = 1\ldots3, i \neq j$. Hence, the trifocal tensor yields an explicit solution to the transfer problem and gives rise to compute motion from more than two images [5].

It is worth mentioning that the epipolar constraint (3.10) can also be regarded as a *necessary* condition relating two projections of a 3-D point **X**: each 2-D motion vector of a rigid object fulfills (3.10), but vice versa, not all vectors satisfying (3.10) represent projected object motion. To obtain a *sufficient* condition, we have to incorporate at least three images and thus employ the trifocal tensor. Then, Equation (3.14) uniquely determines the position in the third image $\mathbf{x}_3$ for given positions in the first $\mathbf{x}_1$ and second $\mathbf{x}_2$ one. Additionally, we should note that both the epipolar matrix and the trifocal tensor form constraints on the image coordinates alone, i.e., they do not require explicit reconstruction of a point **X** in 3-D space. In other words, they decouple 3-D shape from 3-D motion.

3.3 *Approaches to motion estimation*

Numerous approaches toward motion computation are reported in literature. For overview articles on optical flow computation and motion estimation, the interested reader is referred to [6], [7], and [8], respectively. Motion computation techniques, in general, employ the following three models whose particular choice indicates the suitability to a specific application:

1. Observation Model: This model describes how motion and image data interact. Typically, one imposes that some local feature of the image is "almost" invariant along motion trajectories. This feature may be spatial gradient, curvature, color, or just image intensity itself.

2. Motion Field Model: This model restricts the degrees of freedom of motion spatially and possibly also in temporal direction. *Parametric motion field models* impose that the motion field may be described with a finite set of parameters. The best known parametric model assumes 2-D translatoric motion; i.e., it describes the motion field with only 2 parameters $\mathbf{d}(\mathbf{x}) = \mathbf{d}$. Other parametric models augment additional parameters that are often physically meaningful, such as rotation, shear, and zoom. While parametric models impose hard constraints on motion fields, *probabilistic motion field models* impose weak constraints. This can be achieved by modeling motion fields as random fields whose *a priori* probability increases with smoothness. Frequently, Gibbs-Markov random fields are used for this purpose (see [9, 10, 11]).

3. Region Model: The region model describes the region of validity for a particular motion field. Often, regions of fixed size and position (e.g., blocks) are used. Region models that adapt to the shape of independently moving objects allow a more appropriate description and are attractive due to the possibility of assigning and manipulating meaningful object properties. On the other hand, the additional computation of motion-based segmentation increases complexity of motion field computation. Probabilistic models may simultaneously account for motion and region properties. Typically, such models assume regionwise smooth motion with smooth region boundaries and may be formulated as Gibbs-Markov random fields [8].

Next, we discuss some of the important motion computation methods and their underlying models.

3.3.1 *Gradient-based methods*

The best known observation model for motion estimation imposes that intensities are constant along motion trajectories. This assumption holds only approximately in practice, because it ignores phenomena such as occlusions, disocclusions, and intensity variations due to changes in orientation and lighting. The assumption may be expressed for frames at times t and $t+1$ as

$$g(\mathbf{x},t) = g(\mathbf{x}+\mathbf{d}(\mathbf{x},t),t+1), \tag{3.15}$$

where $g(\mathbf{x}, t)$ denotes image intensity and $\mathbf{d} = (d_x, d_y)^{\mathrm{T}}$ is the 2-D motion field in the image plane as defined in Equation (3.3). The constant intensity assumption can also be formulated as differential equation (see [9])

$$\left(\frac{\partial g(\mathbf{x},t)}{\partial x}, \frac{\partial g(\mathbf{x},t)}{\partial y}, \frac{\partial g(\mathbf{x},t)}{\partial t}\right)\begin{pmatrix} d_x \\ d_y \\ 1 \end{pmatrix} = \nabla^{\mathrm{T}} g(\mathbf{x},t)\begin{pmatrix} d_x \\ d_y \\ 1 \end{pmatrix} = 0, \tag{3.16}$$

which is equivalent to the discrete formulation of Equation (3.15) for a short time interval $[t,t+1]$. The above linear equation is underdetermined because it includes two unknowns. It can be seen that only the component of the velocity vector in direction of the spatial brightness gradient can be determined (*aperture effect*). To overcome this, the velocity vector field is constrained to spatial constancy inside some region $\mathcal{R}$ which contains sufficient directionally varying gradient information (corners) to compute both components of the velocity vector. An estimate $\hat{\mathbf{d}}$ of the velocity vector can then be obtained from the solution of the minimization problem

$$\hat{\mathbf{d}} = \arg\min \|\mathbf{e}\|^2; \quad \|\mathbf{e}\|^2 = \sum_{\mathbf{x}\in\mathcal{R}} \left[\nabla^{\mathrm{T}} g(\mathbf{x},t)\begin{pmatrix} \hat{d}_x \\ \hat{d}_y \\ 1 \end{pmatrix}\right]^2. \tag{3.17}$$

This minimization problem can be solved by applying an ordinary least squares approach to the linear equation

$$\begin{pmatrix} \dfrac{\partial g(\mathbf{x}_1)}{\partial x} & \dfrac{\partial g(\mathbf{x}_1)}{\partial y} \\ \vdots & \vdots \\ \dfrac{\partial g(\mathbf{x}_N)}{\partial x} & \dfrac{\partial g(\mathbf{x}_N)}{\partial y} \end{pmatrix}\hat{\mathbf{d}} = \mathbf{G}_{xy}\hat{\mathbf{d}} = -\begin{pmatrix} \dfrac{\partial \hat{g}(\mathbf{x}_1)}{\partial t} \\ \vdots \\ \dfrac{\partial \hat{g}(\mathbf{x}_N)}{\partial t} \end{pmatrix} = -\hat{\mathbf{g}}_t, \tag{3.18}$$

where $N = |\mathcal{R}|$ denotes the number of pixels in the considered region. The LS solution is then given as

$$\hat{\mathbf{d}} = -(\mathbf{G}_{xy}^{\mathrm{T}}\mathbf{G}_{xy})^{-1}\mathbf{G}_{xy}^{\mathrm{T}}\hat{\mathbf{g}}_t. \tag{3.19}$$

The uncertainty of the estimate can be determined considering the deviation $\Delta\hat{\mathbf{d}} = \hat{\mathbf{d}} - \mathbf{d}$ of an observation from its ideal counterpart. A first-order approximation for the deviation between the estimated velocity vector and the true velocity vector can be obtained as

$$\Delta\hat{\mathbf{d}} \approx \left(\frac{\partial\hat{\mathbf{d}}}{\partial\hat{\mathbf{g}}_t}\right)^{\mathrm{T}} \cdot \Delta\hat{\mathbf{g}}_t = -(\mathbf{G}_{xy}^{\mathrm{T}}\mathbf{G}_{xy})^{-1}\mathbf{G}_{xy}^{\mathrm{T}}\Delta\hat{\mathbf{g}}_t. \tag{3.20}$$

From Equation (3.20), a formulation for the covariance matrix is calculated as follows:

$$\mathrm{Cov}(\hat{\mathbf{d}}) = \mathrm{E}\left\{\Delta\hat{\mathbf{d}}\Delta\hat{\mathbf{d}}^{\mathrm{T}}\right\} = (\mathbf{G}_{xy}^{\mathrm{T}}\mathbf{G}_{xy})^{-1}\mathbf{G}_{xy}^{\mathrm{T}}\mathrm{E}\left\{\Delta\hat{\mathbf{g}}_t\Delta\hat{\mathbf{g}}_t^{\mathrm{T}}\right\}\mathbf{G}_{xy}\left((\mathbf{G}_{xy}^{\mathrm{T}}\mathbf{G}_{xy})^{-1}\right)^{\mathrm{T}}. \tag{3.21}$$

Assuming observation vector $\mathbf{g}_t$ is disturbed by stationary zero-mean uncorrelated noise with standard deviation σ, Equation (3.21) reduces to

$$\mathrm{Cov}(\hat{\mathbf{d}}) = \sigma^2(\mathbf{G}_{xy}^{\mathrm{T}}\mathbf{G}_{xy})^{-1} = \frac{\sigma^2}{|\mathbf{G}_{xy}^{\mathrm{T}}\mathbf{G}_{xy}|}\left(\begin{array}{cc} \sum_R\left(\dfrac{\partial g(\mathbf{x})}{\partial y}\right)^2 & -\sum_R\dfrac{\partial g(\mathbf{x})}{\partial x}\dfrac{\partial g(\mathbf{x})}{\partial y} \\[2ex] -\sum_R\dfrac{\partial g(\mathbf{x})}{\partial x}\dfrac{\partial g(\mathbf{x})}{\partial y} & \sum_R\left(\dfrac{\partial g(\mathbf{x})}{\partial x}\right)^2 \end{array}\right).$$

$$\tag{3.22}$$

As expected, variances are large in directions of low spatial gradient and vice versa.

Despite good results that are achieved by this method for small displacements $\hat{\mathbf{d}}$, the underlying model is often criticized for its unequal treatment of spatial and temporal derivatives. The error $\mathbf{e}$ is confined to the observation vector $-\hat{\mathbf{g}}_t$; therefore, the model accumulates all uncertainties in the temporal derivative while it assumes that spatial gradients are free of noise. In practice, however, all gradients will be approximated by filters that work on the sampled image sequence and, hence, uncertainty should be expected in all components of ∇g.

These considerations lead to a *total least squares* approach. For this purpose, Equation (3.16) is rewritten as

$$\nabla g(\mathbf{x},t)\begin{pmatrix} d'_x \\ d'_y \\ d'_t \end{pmatrix} = \nabla g(\mathbf{x},t)\mathbf{d}' = 0, \tag{3.23}$$

where the motion vector **d′** is extended to three components. From this formulation, the desired 2-D motion vector **d** can be readily obtained from the relationship

$$\mathbf{d} = \frac{1}{d_t'}\begin{pmatrix} d_x' \\ d_y' \end{pmatrix}. \tag{3.24}$$

The vector **d′** is obtained by minimizing

$$\hat{\mathbf{d}}' = \arg\min_{\hat{\mathbf{d}}'^{\mathrm{T}}\hat{\mathbf{d}}'=1} \|\mathbf{e}\|^2; \ \|\mathbf{e}\|^2 = \sum_R \left(\nabla^{\mathrm{T}} g(\mathbf{x},t)\hat{\mathbf{d}}'\right)^2 \tag{3.25}$$

subject to the constraint $\hat{\mathbf{d}}'^{\mathrm{T}}\hat{\mathbf{d}}' = 1$, which prevents the trivial solution $\hat{\mathbf{d}}' = 0$. Introducing the *structure tensor*

$$\mathbf{T} = \sum_{\mathcal{R}} \nabla g(\mathbf{x},t)\nabla^{\mathrm{T}} g(\mathbf{x},t) ,$$

(see, for example, [12, 13]), we are left with a constrained minimization problem

$$\|\mathbf{e}\|^2 = \sum_{\mathcal{R}} \hat{\mathbf{d}}'^{\mathrm{T}}\nabla g(\mathbf{x},t)\nabla^{\mathrm{T}} g(\mathbf{x},t)\hat{\mathbf{d}}' = \hat{\mathbf{d}}'^{\mathrm{T}}\mathbf{T}\hat{\mathbf{d}}' , \tag{3.26}$$

which can be reduced via the Lagrange function

$$L(\hat{\mathbf{d}}',\lambda) = \hat{\mathbf{d}}'^{\mathrm{T}}\mathbf{T}\hat{\mathbf{d}}' + \lambda(1-\hat{\mathbf{d}}'^{\mathrm{T}}\hat{\mathbf{d}}') \tag{3.27}$$

to the eigenvalue problem

$$\mathbf{T}\hat{\mathbf{d}}' = \lambda\hat{\mathbf{d}}'. \tag{3.28}$$

Thus, the eigenvector $\hat{\mathbf{d}}'^*$ corresponding to the smallest eigenvalue λ_3 of the structure tensor **T** is the desired solution. The confidence of the measurement can be assessed by rank analysis of the structure tensor as outlined in Table 3.1.

3.3.2   Intensity matching

Matching of intensity regions is the most widely applied method due to its easy implementation and its appropriateness even for large motion.

An image is subdivided into regions (often rectangular blocks) $\mathcal{R}$ of $N = |\mathcal{R}|$ pixels. The region size is chosen small, such that an identical motion vector throughout the block may be assumed. The intensities

Table 3.1 Confidence Analysis of the Motion Estimate Based on the Rank of the Structure Tensor **T**.

rank (**T**)	Uncertainty of Estimation
0	no estimation possible
1	aperture problem: only one component of the velocity vector can be estimated
2	sufficient spatial distribution of the gradients inside the region $\mathcal{R}$; constant motion; eigenvalues correspond to variances
3	no consistency of the motion inside the region $\mathcal{R}$

inside such a block in one image are compared with the values of the region in the second image that is shifted by **d**. Typical measures for dissimilarity d between two regions A,B are symmetrical, i.e., $d(A,B)= d(B,A)$, and nonnegative, i.e., $d(A,B) \geq 0 \quad \forall A,B,$ with $d(A,B)=0 \Leftrightarrow A = B$. Prominent dissimilarity measures d include the sum of squared differences

$$d_{\mathrm{SSD}}(\mathbf{d}) = \frac{1}{N}\sum_{\mathbf{x}\in\mathcal{R}}\left(g_1(\mathbf{x})-g_2(\mathbf{x}+\mathbf{d})\right)^2 \tag{3.29}$$

and the sum of absolute differences

$$d_{\mathrm{SAD}}(\mathbf{d}) = \frac{1}{N}\sum_{\mathbf{x}\in\mathcal{R}}\left|g_1(\mathbf{x})-g_2(\mathbf{x}+\mathbf{d})\right|. \tag{3.30}$$

A well-known similarity measure $s = 1 - d$ is the normalized cross-correlation coefficient

$$s_\rho(\mathbf{d}) = \frac{\displaystyle\sum_{\mathbf{x}\in\mathcal{R}}(g_1(\mathbf{x})-\bar{g}_1)(g_2(\mathbf{x}+\mathbf{d})-\bar{g}_2)}{\sqrt{\displaystyle\sum_{\mathbf{x}\in\mathcal{R}}(g_1(\mathbf{x})-\bar{g}_1)^2\sum_{\mathbf{x}\in\mathcal{R}}(g_2(\mathbf{x}+\mathbf{d})-\bar{g}_2)^2}}. \tag{3.31}$$

where $\bar{g}_1$ and $\bar{g}_2$ denote the mean intensity in the first and second block, respectively. It is worth noting that this measure does not impose constant intensity along motion trajectories but is invariant to contrast scaling a and illumination offsets b. Thus, the observation model Equation (3.15) reads

$$g(\mathbf{x},t) = a\cdot g(\mathbf{x}+\mathbf{d}(\mathbf{x},t),t+1)+b. \tag{3.32}$$

Motion estimation now reduces to computation of the similarity measure s for all motion vector candidates and selection of the motion vector that yields maximum similarity.

Gradient-based as well as intensity-matching methods rely on the assumption that all pixels within a region undergo the same translation. This constraint can be straightforwardly extended to other parametric motion models. However, rarely more than three parameters are chosen as the computational effort increases with the number of motion parameter candidates.

Regardless of the number of parameters, the hard motion constraint does not reflect real motion perfectly. This is especially true for nonrigid motion. This results in a dilemma for the choice of block size. Large blocks may represent real motion only poorly. On the other hand, small blocks may not include sufficient cue for unique identification. In the latter case, motion estimation may become ambiguous in the sense that a large number of similar blocks may exist.

While a similarity measure suffices to compute a motion vector that is optimal in some sense, it is crucial for many tasks to compute an uncertainty measure of the motion estimate. This may be achieved by extending the similarity measure to a probability distribution for the motion vector. The Bayesian paradigm considers both the image data g_2 and the motion field as random fields[2] (see [11]). A suitable distribution of $\mathbf{d}$ for motion estimation is the *posterior*

$$p_{D|G}(\mathbf{d}|g_2) = \frac{p_{G|D}(g_2|\mathbf{d})p_D(\mathbf{d})}{p_G(g_2)}, \tag{3.33}$$

where $p_{G|D}(g_2|\mathbf{d})$ denotes the likelihood of the new image g_2 for given motion $\mathbf{d}$, and the *a priori* distribution $p_D(\mathbf{d})$ represents prior expectations on the estimate. The denominator is a normalization constant for a given image g_2.

We associate a *displaced frame difference* image e to every motion candidate $\mathbf{d}$

$$e(\mathbf{x}) = g_2(\mathbf{x} + \mathbf{d}) - g_1(\mathbf{x}). \tag{3.34}$$

For the "true" motion vector, the constant intensity assumption Equation (3.15) imposes $e = 0$. In practice however, we expect a small but nonzero displaced frame difference e and e will serve as observation error. Assuming that the likelihood for a new image g_2 given the motion $\mathbf{d}$ and the deterministic first image g_1 only depends on the displaced frame difference, and distributing our prior expectations uniformly over the set $\mathcal{D}$ of all possible motion vectors, Equation (3.33) can be rewritten as

$$p_{D|G}(\mathbf{d}|g_2) = \frac{p_E(e)}{\displaystyle\sum_{\mathbf{d} \in \mathcal{D}} p_E(e)}, \tag{3.35}$$

---

[2] It is sufficient to model the first image g_1 deterministically.

where the denominator normalizes the *posterior* to unity.

The most common model for the observation error distribution is a zero-mean white Gaussian (see, e.g., [14, 15]). Grouping the observation error in the considered region $\mathcal{R}$ to a vector $\mathbf{e} = (e(\mathbf{x}_1), e(\mathbf{x}_2), \ldots, e(\mathbf{x}_N))^T$, we can write

$$p_E(\mathbf{e}) = \left(1 / \sqrt{2\pi\sigma_e^2}\right)^N \exp\left[-\sum_{i=1}^N e(\mathbf{x}_i)^2 / 2\sigma_e^2\right], \tag{3.36}$$

where the variance σ_e^2 comprises modeling and observation noise. Insertion of this equation into Equation (3.35) readily yields the desired *posterior*

$$p_{D|G}(\mathbf{d}|g_2) = \frac{\exp\left[-\sum_{i=1}^N e(\mathbf{x}_i)^2 / 2\sigma_e^2\right]}{\sum_{d \in \mathcal{D}} \exp\left[-\sum_{i=1}^N e(\mathbf{x}_i)^2 / 2\sigma_e^2\right]}, \tag{3.37}$$

depending on $\mathbf{d}$ through Equation (3.34). In the light of Equation (3.37), the sum of squared differences matching criterion, Equation (3.30), yields the *maximum likelihood (ML)* estimate of the 2-D motion vector, i.e., the displacement vector maximizing Equation (3.37). Likewise, the sum of absolute differences matching criterion, Equation (3.30), yields the ML estimate for a Laplacian observation model.

Alternatively, one could also choose to determine the *linear minimum variance estimate* of $\mathbf{d}$ as the mean (rather than the mode) of the discrete probability density function (3.37)

$$\bar{\mathbf{d}} = E\{\mathbf{d}\} = \sum_{d \in \mathcal{D}} \mathbf{d} \cdot p_{D|G}(\mathbf{d}|g_2). \tag{3.38}$$

Equation (3.37) elegantly allows access to the uncertainty of the estimate through its covariance matrix

$$E\{(\mathbf{d} - \bar{\mathbf{d}})(\mathbf{d} - \bar{\mathbf{d}})^T\} = \sum_{d \in \mathcal{D}} (\mathbf{d} - \bar{\mathbf{d}})(\mathbf{d} - \bar{\mathbf{d}})^T p_{D|G}(\mathbf{d}|g_2)$$

$$= \begin{pmatrix} \sigma_{d_x}^2 & \rho \sigma_{d_x} \sigma_{d_y} \\ \rho \sigma_{d_x} \sigma_{d_y} & \sigma_{d_y}^2 \end{pmatrix}, \tag{3.39}$$

where $\sigma_{d_x}^2$ and $\sigma_{d_y}^2$ denote the variance of the motion estimate in the *x*- and *y*-direction, respectively, and ρ is the correlation coefficient. The eigenvectors of matrix (3.39) incorporate information about the directional uncertainty of the displacement error. The eigenvector associated with the largest eigenvalue of (3.39) gives the direction of greatest uncertainty, whereas the eigenvector of the smallest eigenvalue points in the direction of highest confidence in the motion estimate. This is illustrated as confidence ellipses (i.e., contours of constant pdf) in the image plane in Figure 3.4.

Finally, the probabilistic model allows one to perform a *goodness-of-fit (GOF) test*, such as the χ^2-test. The GOF measure is extracted from the cumulative χ^2-distribution with N-2 degrees of freedom with

$$\chi^2 = \sum_{i=1}^{N} \hat{e}(\mathbf{x}_i)^2 / \sigma_e^2 .$$

An unlikely large residuum $\hat{\mathbf{e}}$ indicates a model violation; i.e., no motion vector explains the temporal image changes satisfactorily. This may be the

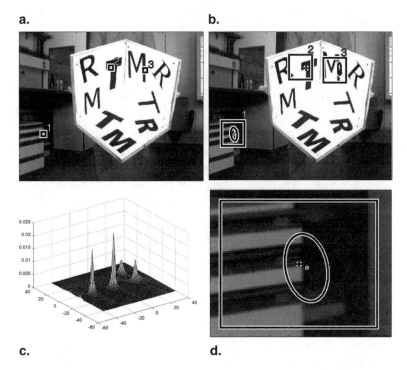

a.   **b.**

c.   **d.**

Figure 3.4 (a) and (b): two consecutive image frames with three matched image points. The confidence ellipses are also plotted for each match; (c): the probability density function for match 1 with ambiguous local maxima; (d) results for match 1 and the corresponding search ROI in the second image. Note that the centroid of the pdf ('.') does not coincide with the mode of the pdf ('+').

case at occlusions, motion that cannot be described with the chosen motion model or an inappropriate segmentation. Thus, the GOF test provides valuable information for many applications.

3.3.3   Feature matching

All methods considered so far compute motion through comparing intensities within an image region. This comparison implies a motion model that constrains the variation of motion vectors spatially and temporally. A natural choice for such a motion model does not exist. Furthermore, the observation model of constant intensities along motion trajectories may be significantly violated in some imagery.

It is often more appropriate to assume invariance of some feature at prominent points in an image rather than imposing invariant intensity within a whole region. Furthermore, low-level cues of objects like corners and edges are often better represented by feature vectors than by intensity values within a region.

Numerous features are proposed for motion computation in literature. These include edges, corners, local entropy, and local contrast. In the sequel of this section, we will focus on a concise class of features that has proven powerful for motion computation, namely the shape context.

3.3.3.1   Shape context matching

Shape context matching (SCM) was originally proposed for shape recognition [16]. It relies on the assumption that the shape of an object can be appropriately represented by its internal and external contours. For this purpose, consider the set of vectors originating from a point x_1 on a contour to all other sample points on the shape. These vectors express the configuration of the entire shape relative to the selected reference point. The points corresponding to these vectors can be obtained as locations of edge pixels as found by an edge detector yielding a set $\mathcal{B} = \{x_1, \ldots, x_n\}$, $x_i \in R^2$ of n points. This point set of subsampled edges is not necessarily limited to consist of curvature extremes or inflection points. Usually, $\mathcal{B}$ is obtained by subsampling the edge image uniformly. As more samples are used, a better approximation of the underlying shape can be obtained. Since the sample points are drawn independently from the two shapes, the shape context vector is subject to jitter noise. This error can be reduced by approximating the sampled shape points with a parametric curve model prior to matching [17].

The shape context descriptor represents the coarse distribution of the shape (points) with respect to a particular point on the shape. Figure 3.5 illustrates the generation of the shape context descriptor. The distribution of relative positions yields a robust, compact, and highly discriminative descriptor. The vectors of relative edge positions are quantized into bins as depicted in Figure 3.5(c). For a point x_i on the shape, we compute a coarse histogram h_i of the relative coordinates of the remaining $n-1$ points x_j

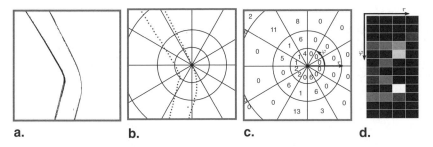

Figure 3.5 Computation of the shape context for one edge point: (a) binary edge image; (b) subsampled edges with overlaid log-polar coordinate system; (c) log-polar representation of two-dimensional histogram of (b); (d) two-dimensional histogram in matrix form (shape context descriptor).

$$h_i(k) = \left| \{ j \neq i : (\mathbf{x}_j - \mathbf{x}_i) \in \text{bin}(k) \} \right|. \tag{3.40}$$

This 2-D position histogram associated with $\mathbf{x}_i$ is called the shape context of this point. To make the descriptor more sensitive to positions of nearby sample points, bins that are uniform in log-polar space are used as shown in Figure 3.5(c) and (d).

Shape correspondences are found through minimization of the distance

$$C_{ij} = C(\mathbf{x}_{1,i}, \mathbf{x}_{2,j}) = \frac{1}{2} \sum_{k=1}^{K} \frac{(h_{1,i}(k) - h_{2,j}(k))^2}{h_{1,i}(k) + h_{2,j}(k)} \tag{3.41}$$

between the shape context histograms $h_{1,i}(k)$ and $h_{2,j}(k)$ of two points $\mathbf{x}_{1,i}$ and $\mathbf{x}_{2,j}$ on shapes in the first and second image, respectively.

Shape context matching helps to incorporate global image properties in the matching process. Additionally, it does not depend on invariance of pixel intensities but, instead, depends only on the invariance of relative positions of the feature points. Therefore, shape context matching yields a substantial amount of robustness against noise in the intensities along the motion trajectories and against spatially varying motion.

3.4 Applications

3.4.1 Segmentation-based image sequence coding

Bandwidth-efficient transmission of image sequences is the key to numerous applications in video communications and video data storage. While image sequences already show strong statistical bindings of intensities in spatial and temporal direction, bindings directed along motion trajectories are even significantly stronger. The savings in code length gained through exploitation of these bindings were found to overcompensate for the expenditure in code length for motion encoding. Thus, it comes as no surprise that the broad majority of source coding techniques incorporate explicit motion information. All coding standards involve a combination of motion-compensated prediction and some residual encoding technique [18]. This *hybrid coding concept* is depicted in Figure 3.6.

Methods proposed for the estimation and encoding of motion information, however, still exhibit broad variety. Typically, motion is described using some parametric model that accounts for the dynamics of typical scenes. Both 2-D and 3-D representations of motion have been proposed (see, e.g., [19, 20]). The region model employed for motion validity has strong impact on the functionality available as well as on the computational effort required.

In the simplest case, so-called *global motion* is encoded; i.e., the validity segment is the whole frame. At the other extreme, *dense motion* incorporates one motion vector per pixel, e.g., [11]. An approach that is appealing for its reasonable compromise between computational load and appropriateness of motion description is *block-based motion* encoding, where the image is segmented into rectangular blocks. One set of motion parameters is then encoded per block. Blocks may be of fixed or scene-adapted size; in the latter case, the individual block partitioning has to be encoded as shape information. *Region-based motion* accounts for the fact that motion fields are typically segmentwise smooth, where the 2-D motion-based segmentation ideally corresponds to the projection area of individually moving objects, e.g., [21, 22]. In this concept, the 2-D shape of the individual segments has to accompany the motion parameter sets. The segmentation not only is meant to enhance efficiency of motion field encoding but also provides meaningful information by itself. This information enables applications such as segmentwise scene manipulation. In the very extreme, *object-based*

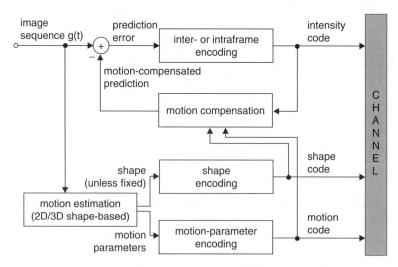

Figure 3.6 Block diagram of a hybrid cooler.

motion and shape information may correspond to 3-D models for objects in the 3-D real world. Such codes offer the highest degree of physical meaningfulness and enable object-based 3-D scene manipulation with a broad spectrum of prospective applications. Although they also show large theoretical potential to increase encoding efficiency to a large extent, likewise small benefits have been reported so far for real encoders, and estimation and encoding strategies of 3-D motion and shape are furthest from their reliable and practical realization.

The negative logarithm of the Bayesian formulation of Equation (3.33) allows an interesting interpretation

$$-\log p_G(g_2) = -\log p_{G|D}(g_2|\mathbf{d}) - \log p_D(\mathbf{d}) + \log p_{D|G}(\mathbf{d}|g_2). \qquad (3.42)$$

We recall that the negative logarithm of probability is the content of information as defined in the fundamental article by Shannon [23]; i.e., it is just the code length of an optimal lossless encoder. Hence, the code length to transmit a new image g_2 is the sum of the code length of the new image for known motion, i.e., the prediction error image, and the code length for motion itself. Assuming any deterministic motion computation method, the last term becomes zero. Therefore, a hybrid coder as depicted in Figure 3.6 must find a balance between code length for the prediction error and code length for the motion information. This has led to gain/cost controlled motion estimation methods, where the estimator trades the error of motion compensated prediction against code length of motion field encoding [24, 25]. It is worth noting that these coding methods are closely related to minimum description length (MDL) estimators.

The above discussion shows that motion field encoding needs to be carefully selected. A reasonable distortion measure for the motion field in the context of coding is the squared error

$$d^2 = \sum_{\mathbf{x} \in \mathcal{R}} \left[g_2(\mathbf{x}) - g_1(\mathbf{x} - \hat{\mathbf{d}}(\mathbf{x})) \right]^2, \qquad (3.43)$$

where $\hat{\mathbf{d}}(\mathbf{x})$ denotes the encoded motion field. This measure is consistent with the dissimilarity measure of Equation (3.29) that is frequently used for motion computation. Approximation of the nonlinear image intensity pattern by a first-order Taylor series eventually yields

$$d^2 = \sum_{\mathbf{x} \in \mathcal{R}} \left[\nabla^T g_2(\mathbf{x}) \Delta \mathbf{d} \right]^2, \qquad (3.44)$$

where ∇g denotes spatial intensity gradient and $\Delta \mathbf{d} = \mathbf{d} - \hat{\mathbf{d}}$ is the deviation of the encoded motion from the "true" motion fulfilling the assumption of constant intensities along motion trajectories. Hence, the distortion measure weights the deviation $\Delta\mathbf{d}$ of a motion vector with the spatial gradient. From the viewpoint of estimation uncertainty, which typically exhibits a standard deviation that is inverse proportional to spatial intensity gradient as, e.g., in Equation (3.21), Equation (3.44) allows the following interesting interpretation: to achieve uniform image quality, an estimate of $\mathbf{d}$ should be encoded with an accuracy that is proportional to its standard deviation.

Encoding the motion field

$$\hat{\mathbf{d}}_{\mathcal{R}} = \left(\hat{\mathbf{d}}(\mathbf{x}_1), \hat{\mathbf{d}}(\mathbf{x}_2), \ldots, \hat{\mathbf{d}}(\mathbf{x}_N) \right)$$

in the considered region $\mathcal{R}$ as a linear combination of $M \leq N$ basis functions $\mathbf{u}_{\mathcal{R},i} = \left(u_i(\mathbf{x}_1), u_i(\mathbf{x}_2), \ldots, u_i(\mathbf{x}_N) \right)$, i.e.,

$$\hat{\mathbf{d}}_{\mathcal{R}} = \sum_{i=1}^{M} \hat{c}_i \mathbf{u}_{\mathcal{R},i}, \tag{3.45}$$

the task of motion field encoding reduces to identification and encoding of a suitable set of M coefficients $\hat{c}_i \in R^2$. This encoding scheme can be regarded as a region-based linear transform of the motion field that is separable in horizontal and vertical motion. For general choice of transform bases and a sufficient number of coefficients, an arbitrary motion field may be losslessly encoded. Favorable choices for $\mathbf{u}_{\mathcal{R},i}$ yield a high energy compaction toward low order coefficients, i.e., allow a good approximation of motion fields with a small number of nonzero coefficients. Polynomial [22] or DCT bases are particularly appealing for this purpose. Equations (3.44) and (3.45) define a weighted squared distortion measure and a linear observation model for $\hat{c}_i$; i.e., the coefficients can be readily computed as linear weighted least squares (WLS) estimates.

Figure 3.7 depicts two consecutive segmentation-based dense motion fields computed with a Bayesian estimator that employs a Gibbs-Markov model as proposed in [26].

The model favors segmentwise smooth motion, smooth segment boundaries, and temporal continous motion along motion trajectories.

The left side of Figure 3.8 shows the motion-compensated prediction for half-pixel accurate 16×16 block matching minimizing the SSD distortion measure of Equation (3.29). This is compared to the motion-compensated prediction employing segmentation-based motion as of the figure on the right side.

The bottom images show the motion-compensated frame difference for the two motion estimates. The superiority of segmentation-based motion-compensated prediction is clearly visible, from an objective measure of difference energy as well as by subjective impression.

Figure 3.7 Salesman frames 125 and 129 with superimposed dense segmenta-tion-based motion (subsampled for visualization).

Figure 3.8 Motion compensated prediction of salesman 125 for different motion es-timates. Left: half-pixel accurate block-based motion; Right: segmentation-based dense motion as depicted in Figure 3.7. The bottom shows motion-compensated frame difference images.

Figure 3.9 relates the number of motion coefficients that is needed for block-based motion fields to the number of motion coefficients computed with a WLS algorithm according to Equations (3.44) and (3.45).

Block-based motion requires a block size less than 4×4 to match the quality of segmentation-based motion encoding. The savings in data rate for identical PSNR overcompensate for the necessity of segmentation encoding. The objec-tive and subjective gain in image quality and the ability of segmentwise image

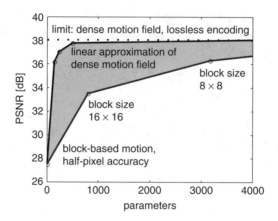

Figure 3.9 PSNR of motion-compensated prediction for block-based and segmentation-based motion (without contour encoding).

manipulation must, however, be paid for by a significant increase of complexity for image description and motion estimation.

3.4.2 Velocity sensor

Numerous applications require accurate knowledge of the physical velocity of a technical object. In production, process monitoring, e.g., precise velocity measurements of uniformly moving production goods, is essential during all stages of the production cycle. Another application of 3-D motion sensing is monitoring of slippage and tension of v-belts in drives. Here, knowledge of velocities enables supervision of strain and forces. As a last example, the measurement of speed over ground of air or ground vehicles is an important issue. Optical sensing of velocities offers many advantages over alternative methods as they are

1. free of contact: optical sensors are easy to install and do not influence the measured quantity.
2. self assessing: optical sensors can assess their actual accuracy (e.g., in terms of a covariance) and reliability. Furthermore, optical sensors may detect and signal degradations in their pose and orientation as well as sensor contamination.

Clearly in all the above mentioned applications, the desired measurement is the physical, i.e., "true" motion, rather than optical flow, i.e., "apparent" motion. Each measurement must be accompanied by a measure for its uncertainty to enable a meaningful interpretation or appropriate action by a subsequent controller.

Figure 3.10 depicts a simple yet practical configuration for such measurements. A camera observes a technical surface moving with velocity **v**. Surface and image planes are parallel. In general, 3-D object motion has six

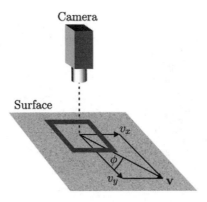

Figure 3.10 Camera observing a moving surface.

degrees of freedom, namely 3-D rotation and 3-D translation. Often, system configuration constrains these to a few unknowns. In the example given, one is interested in the two translatoric components of instantaneous velocity in the surface plane **v**.

For a constant distance between camera and technical surface, $Z = Z_0$, the perspective projection of Equation (3.4) reduces to a scaling of the X- and Y-coordinates:

$$\mathbf{x}(t) = \pi(\mathbf{X}(t)) = \frac{1}{Z_0}\begin{pmatrix} X(t) \\ Y(t) \end{pmatrix} \tag{3.46}$$

Denoting the time between two consecutively acquired images $g_1(\mathbf{x})$ and $g_2(\mathbf{x})$ by Δt, Equation (3.15) reads

$$g_1(\mathbf{x}) = g_2(\mathbf{x} + \mathbf{v}\Delta t), \tag{3.47}$$

i.e., the motion vector

$$\mathbf{d} = \mathbf{v}\Delta t \tag{3.48}$$

is constant for all **x**.

Figure 3.11 depicts two frames from a synthetic image sequence of a moving textured surface for the configuration of Figure 3.10. The shift between the images is 20 pixels in the x-direction and 10 pixels in the y-direction, respectively. A pair of corresponding blocks is marked in the motion field in the right of Figure 3.11(c).

The velocity vector can be computed by any of the methods of Section 3.3. Intensity matching methods employing block-shaped regions and an appropriate similarity measure offer themselves for their low complexity

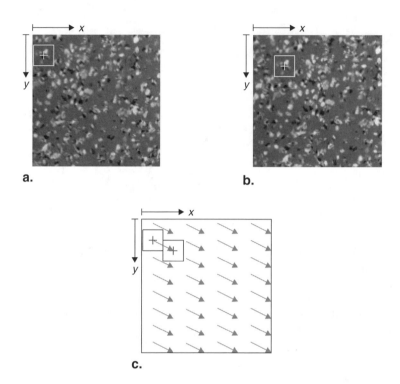

Figure 3.11 Images of a real technical surface with synthetic motion: (a) $t = t_1$; (b) $t = t_1 + \Delta t$; (c) synthetic homogeneous motion field.

and robustness. We employ the normalized cross-correlation coefficient of Equation (3.31) for its invariance against varying illumination.

For the case of spatially constant motion, matching of a single block would already allow computation of the velocity vector $\mathbf{v}$ via Equation (3.48).

However, matching may yield unreliable results in regions of poor or ambiguous texture. In order to increase reliability and accuracy of the computed velocity $\hat{\mathbf{v}}$, we compute velocity $\hat{\mathbf{v}}_i$ for several blocks $i = 1, \ldots, M$. A thorough analysis and fusion of these estimates necessitates the association of each motion estimate with its uncertainty. Formulation of uncertainty in the form of a covariance matrix $\mathrm{Cov}(\hat{\mathbf{v}}_i)$, like defined in Equation (3.39), provides important information about the quality of each estimate. Large standard deviations of both components indicate poorly structured regions, where no reliable computation can be expected. One large and one small eigenvalue of the covariance matrix occur in regions of directed structure, such as an edge. There, the velocity component directed along the edge incorporates large uncertainty, whereas the velocity component perpendicular to the edge can be accurately estimated. Finally, two small standard deviations can be expected in highly structured regions (see Figure 3.4).

Fusion of the velocities computed for the individual blocks may be conducted by a linear least-square estimation with

$$\hat{\mathbf{v}} = \left[\sum_{i=1}^{M} \mathrm{Cov}^{-1}(\hat{\mathbf{v}}_i) \right]^{-1} \sum_{i=1}^{M} \mathrm{Cov}^{-1}(\hat{\mathbf{v}}_i) \cdot \hat{\mathbf{v}}_i, \tag{3.49}$$

where each block velocity is weighted with its inverse covariance.

This technique is illustrated in the example of a synthetic sequence resembling a moving textured surface. The "true" motion from one image frame to the next is $\mathbf{d} = (6,4)^{\mathrm{T}}$ pixels. The image frames have been corrupted by additive white Gaussian noise of strong power spectral density. Due to stationarity of the surface texture, the individual velocities are computed with identical standard deviation. Figure 3.12 depicts two frames of the sequence with two corresponding points superimposed.

In two experiments, velocity has been computed from $M=3$ and $M=6$ block estimates, respectively. Figure 3.13 depicts the histograms for the magnitude $|\hat{\mathbf{v}}|$ and for the two velocity components $\hat{v}_x$ and $\hat{v}_y$ acquired from a long image sequence.

The experimental findings of Figure 3.13 are in good agreement with the theoretical reduction of the standard deviation of the velocity component measurement proportial to $\sqrt{M}^{-1}$. Clearly, the measurements with 6 blocks spread less as compared to the measurements with 3 blocks.

Table 3.2 quantifies the standard deviations found in the experiments. The results show that methods based on matching block-shaped regions allow a precise measurement of the velocity of various technical surfaces.

As an extension, motion computation can be augmented by gradient-based methods to enhance the resolution of the individual velocity estimates. Possible outliers can be accounted for through adoption of a robust fusion technique (see, e.g., [27]).

3.4.3 Image alignment

Another typical problem in which motion computation comes into play is the assessment of dissimilarities between two objects in the real world. Such

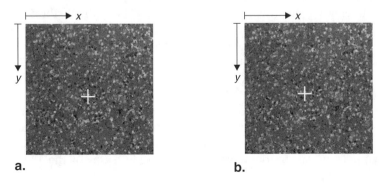

a. **b.**

Figure 3.12 Two subsequent frames of a moving surface. The corresponding points indicate the "true" motion.

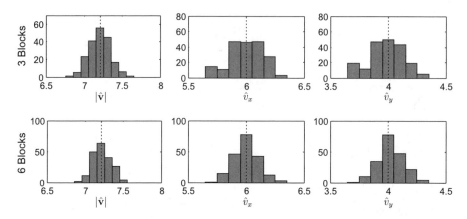

Figure 3.13 Histograms of measured velocities using $M = 3$ and $M = 6$ block velocities, respectively. The "true" value is marked with a dashed line. Velocities are in pixels per frame.

Table 3.2 Standard Deviations of the Estimated Velocities

Matched blocks	σ_v	σ_{v_x}	σ_{v_y}
3	0.141	0.142	0.143
6	0.118	0.107	0.108

tasks frequently occur during production, in which a product has to be compared against a master piece or a CAD model during all stages of the production process. In order to allow a direct comparison of local features, the image data first needs to be aligned, i.e., compensated for different camera pose and orientation.

In a concise application, two gradient images $\mathbf{m}_1(\mathbf{x})$ and $\mathbf{m}_2(\mathbf{x})$, generated by a high-precision deflectometric measuring system [28], of an area of a painted door of a passenger car are to be compared. Since car body components involve very smooth surfaces that do not lend themselves to a reliable intensity matching, feature matching has been the method of choice in this application. Features are extracted at positions of significant curvature, and the computed 2-D motion vectors at these positions are used to estimate the parameters of a transform that aligns the two objects.

3.4.3.1 Feature extraction

For the envisaged purpose, appropriate features should be invariant against 3-D translation and rotation. Such features are, e.g., the local normal curvatures κ_1 and κ_2 of a point on the object in the direction of the main local curvature. They can be obtained from the eigenvalues of the Weingarten mapping matrix [29]. The Weingarten mapping matrix can be approximated using a simplified form of the structure tensor [12, 29]

$$T(x) = \int_{\mathcal{R}} w(x - x') \cdot (m(x')m(x')^{T})dx'$$

$$\approx \begin{pmatrix} \sum_{\mathcal{R}} m_x^2 & \sum_{\mathcal{R}} m_x m_y \\ \sum_{\mathcal{R}} m_x m_y & \sum_{\mathcal{R}} m_y^2 \end{pmatrix} \qquad (3.50)$$

where w denotes an appropriate window function.

The main curvature values can be calculated from a quadratic equation as

$$\kappa_{1,2} = \frac{1}{2}\operatorname{trace}(T(x)) \pm \frac{1}{2}\sqrt{\operatorname{trace}^2(T(x)) - 4|T(x)|}. \qquad (3.51)$$

For simple surfaces the directions of the main curvature are orthogonal [29]. For the comparison with reference data, areas with distinct curvature values are favorable. For this reason, a point is selected as a feature point if the second eigenvalue is larger than a threshold t_1 and the quotient of the main curvature values lies in the range

$$t_2 < \frac{\kappa_1}{\kappa_2} < t_3.$$

This ensures that only areas with a distinct direction of the local gradient change are used for matching while sharp edges may be excluded. Depending on the object under inspection, it may be necessary to exclude sharp object borders because they are often produced with a lower quality (e.g., hemmed edges).

The feature points calculated from the curvature images are used to create binary border images suitable for the shape context matching described in Subsection 3.3.3.

The image registration ensures that the deviation of the object from the reference data is calculated from corresponding surface areas only. In the case of a misalignment or no alignment at all, even any wanted surface curvature would contribute to the deviation. Therefore, all curved areas would be interpreted as faulty.

The deflectometric data used in this example cannot be converted into three-dimensional coordinates easily. On the other hand, the motion between the two data sets is rather small. For this reason, a two-dimensional affine transform is used for the alignment

$$\mathbf{x}'_{i,\text{obj}} = \begin{pmatrix} x'_{i,\text{obj},1} \\ x'_{i,\text{obj},2} \\ 1 \end{pmatrix} = \mathbf{A} \cdot \mathbf{x}_{i,\text{ref}} = \begin{pmatrix} a_1 & a_2 & a_3 \\ a_4 & a_5 & a_6 \\ 0 & 0 & 1 \end{pmatrix} \cdot \begin{pmatrix} x_{i,\text{ref},1} \\ x_{i,\text{ref},2} \\ 1 \end{pmatrix} ; i = 1 \ldots m, \qquad (3.52)$$

where m denotes the number of matched correspondences.

For the determination of the six parameters $a_1 \cdots a_6$ of the affine transform three general point correspondences are sufficient. To obtain a reliable estimate of the parameters, a least median of squares (LMedS) estimator is used that considers typically 100–1000 matches. The LMedS belongs to the family of robust estimation techniques. A subset of at least three point correspondences is chosen randomly and used to calculate the affine parameters with a least-square estimator. This procedure is repeated n times ($n > 100$). For each of the resulting n estimates of the matrix $\mathbf{A}$, the median of the squared distances between measurements and adjusted observations is calculated

$$e_i = \text{med}\left((\mathbf{A}_i \cdot \mathbf{x}_{j,\text{ref}} - \mathbf{x}_{j,\text{obj}})^2 \right) \quad \text{with} \quad j = 1 \ldots m.$$

The matrix $\mathbf{A}_i^*$ yielding the smallest value of e_i is considered the best estimate for the matrix $\mathbf{A}$. Then, the actual image alignment is performed applying the affine transform $\mathbf{A}_i^*$ to the reference image. In a final postprocessing step, the difference image is high pass filtered and slight inconsistencies at the object borders are clipped. An example of the evaluation of a car body part using image alignment is shown in Figure 3.14.

3.4.4 Object detection

Object detection and scene segmentation are important problems encountered in a variety of applications. We have pointed out this fact before; e.g., the segmentation of a scene into layers of different objects can improve the efficiency of video coding and enables object-based manipulation of imagery (cf. region-based motion, Subsection 3.4.1). Furthermore, the detection of independently moving objects in the vicinity of a mobile observer is a vital skill in environmental perception. In the field of robotics, in navigation, and in modern driver assistance systems, environmental perception is the key to increase safety and comfort. Visual motion provides a strong clue to scene segmentation and object detection. For example, camouflaged objects can often only be detected as they begin to move. As will be outlined below, 2-D motion is also a general characteristic to separate independent scene elements, since the only constraint it imposes on the different scene elements is rigidity. In particular, no prior assumptions about the shape of all possible objects have to be made. For these reasons, object detection can be approached by interpreting visual motion data, i.e., by clustering image points that seem to belong to the same rigid object.

A straightforward and somewhat naive approach to scene segmentation is to group points with similar 2-D motion or to establish object boundaries

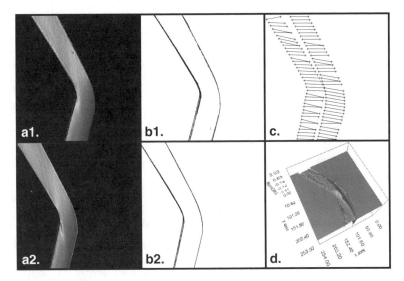

Figure 3.14   (a1) and (a2): original data; (b1) and (b2): binary edge images of (a1) and (a2); (c): motion vectors; (d): difference image after alignment.

that correspond to discontinuities in the 2-D motion field. These methods will work only in scenes with simple geometry. However, in a general environment with objects of discontinuous depth and with motion that includes rotation, these approaches are destined to fail.

In Section 3.2 we have presented a general constraint on 2-D motion that is valid for each point of any rigid object: Each rigid scene element is subject to a 3-D motion transformation $(\mathbf{R}, \mathbf{t})$ following Equation (3.6). As pointed out earlier, each of these transformations constitutes an independent epipolar constraint (3.10)

$$\begin{pmatrix} x \\ y \\ 1 \end{pmatrix}^{\mathrm{T}} \mathbf{E} \begin{pmatrix} x + d_x \\ y + d_y \\ 1 \end{pmatrix} = 0, \tag{3.53}$$

where $(x, y)^{\mathrm{T}}$ denotes the coordinates of the image point and $(d_x, d_y)^{\mathrm{T}}$ the associated 2-D motion vector.

Based on these findings, an algorithm could be deduced to group all image features that are consistent with one individual constraint (3.53). In theory, such an approach could handle any scene containing rigid objects, given the 2-D motion field is dense enough to obtain a reliable estimate of the essential matrix of Equation (3.53). Torr et al. [30] implemented a scene segmentation based on the presented motion constraints.[3] They made elaborate use of robust estimation techniques to recover the epipolar matrices of all objects within the scene and to cluster all points consistent with the same geometric constraint.

It is easy to understand that the epipolar constraint (3.10) only forms a necessary condition for two image points to correspond. The impact of this property on motion segmentation can be illustrated by a simple example: Assume a scene with a static camera and two rigid objects. One is translating with a 3-D motion vector $\mathbf{t}$, the other with a different vector $\mu\mathbf{t}, \mu \neq 0, \mu \neq 1$. Based on (3.10) alone, it will not be possible to distinguish both objects since all image features of the first object will fulfill the epipolar constraint of the second object and vice versa. This ambiguity can only be resolved by introducing a third camera view; thus, Torr et al. propose a segmentation based on the trilinear constraint that relates corresponding points in three different camera images (see [3]).

Here we will describe a different example that uses a sequence of stereo images. The algorithm explicitly recovers 3-D structure and 3-D motion of observed scene elements using a Kalman filter and exploits this information to distinguish between independently moving objects. A more detailed description of this approach can be found in [32].

The principle underlying this approach is tracking a rigid point cloud in a stereo image sequence. We assume that the 3-D structure of the tracked object is represented by 3-D vectors $\mathbf{X}_0^i, i = 1 \ldots N$, which contain the 3-D coordinates of N object points at an initial time instant $t = 0$. If the observed object undergoes rigid motion, the current 3-D positions $\mathbf{X}_c^i$ of the point cloud at a time $t \neq 0$ are obtained by

$$\mathbf{X}_c^i(t) = \mathbf{Rot}(\mathbf{\Omega}(t))\mathbf{X}_0^i(t) + \mathbf{T}(t), \tag{3.54}$$

where $\mathbf{\Omega}(t)$ and $\mathbf{T}(t)$ denote the *cumulated* rotation and translation within time $[0,t]$. $\mathbf{\Omega}(t)$ should be given as Euler angles, and we denote the mapping of a rotation vector to a rotation matrix by $\mathbf{Rot}(\mathbf{\Omega})$.

Under the further assumption that the object moves with instantaneous angular velocity ω and translational velocity $\mathbf{v}$, we can formulate a dynamics model as follows

$$\mathbf{X}_0(t+1) = \mathbf{X}_0(t)$$

$$\mathbf{\Omega}(t+1) = \omega(t) + \mathbf{\Omega}(t)$$

$$\mathbf{T}(t+1) = \mathbf{Rot}(\omega(t)\mathbf{T}(t)) + \mathbf{v}(t) \tag{3.55}$$

$$\omega(t+1) = \omega(t) + \mathbf{n}_\omega$$

$$\mathbf{v}(t+1) = \mathbf{v}(t) + \mathbf{n}_v.$$

---

[3] The algorithm presented in [30] is designed to work on uncalibrated images; i.e., instead of epipolar constraints as in Equation (3.10), they use the fundamental matrices (see [31]). These matrices comprise 3-D motion as well as the (intrinsic) projection parameters of the pinhole camera model. However, in many applications the intrinsic camera parameters are known or at least constant, such that simplified version of their approach will be sufficient.

The system state of Equation (3.55) comprises the constant but initially unknown 3-D structure $\mathbf{X}_0^i$, the instantaneous velocities ω and $\mathbf{v}$, and the cumulated 3-D motion Ω and $\mathbf{T}$. The angular acceleration $\mathbf{n}_\omega$ and the translational acceleration $\mathbf{n}_v$ are modeled as white Gaussian noise. As shown in [33], the above system can be improved by decomposing $\mathbf{X}_0^i$ into its projection $\mathbf{y}_0^i = \pi(\mathbf{X}_0^i)$ and its associated depth $\rho_0^i = X_{0,z}^i$. Thus, uncertainty in the projection of $\mathbf{X}_0^i$ is separated from uncertainty in its distance. In fact, $\mathbf{y}_0^i$ is exactly known and does not have to be estimated by the following Kalman filter.

From stereo image sequences, we can obtain two different types of measurements: 2-D motion between two frames of one camera and stereo disparity between simultaneously acquired left and right camera images. First, the 2-D motion $\mathbf{d}^i$ of an object point $\mathbf{X}_c^i$ between consecutive frames of the right camera is given by

$$\mathbf{d}^i = \pi(X_{c,z}^i(t+1)) - \pi(X_{c,z}^i(t)) \qquad (3.56)$$

where, using the decomposition of $\mathbf{X}_0^i$ as described above, the current position is given by

$$X_c^i(t) = \mathbf{Rot}(\Omega(t)) \cdot \pi^{-1}(\mathbf{y}_0{}^i(t), \rho_0^i(t)) + \mathbf{T}(t).$$

Second, stereo information can be exploited. Since in many applications the epipolar geometry of the stereo cameras is known from calibration, we are able to preprocess the images such that the epipolar lines in both images coincide with the horizontal scan lines of the two camera images.[4] In this simple case, the disparity Δ^i is defined as the difference between the horizontal coordinates of corresponding image points in stereo frames. Δ^i is related to the depth of the observed object point $\mathbf{X}_c^i$ by a simple triangulation formula (see [31])

$$\Delta^i(t) = \frac{B}{X_{c,z}^i(t)} = \frac{B}{\rho_0^i(t) + T_z(t)}, \qquad (3.57)$$

where B denotes the distance between the left and right cameras. Equations (3.56) and (3.57) constitute the measurement equations. Together with the system Equations (3.55) they specify the complete state space model. From this point of view, the described algorithm can also be regarded as a fusion of 2-D motion and stereo disparity.

Computation of the observation vector with both 2-D correspondences — 2-D motion between two consecutive right image frames and disparity

---

[4] This process is known as rectification and is further described in [35].

between left and right stereo image — is conducted by a two-step procedure:

1. Block matching is used to retrieve an initial estimate of the image displacement. To obtain a dense displacement field, matching with adaptive block sizes was implemented. Furthermore, quality measures based on the matching residuals and on local image texture have been established.
2. The previous estimate is refined on subpixel level (i.e., within a range of $\pm 0,5$ pixels). For 2-D motion, a total least-squares algorithm as described in Section 3.31 has been applied.

The system and measurement equations allow the recovery of the 3-D structure and the 3-D motion of the tracked object in a standard Kalman filter. Since Equations (3.55), (3.56), and (3.57) are nonlinear, an iterated extended Kalman filter (IEKF), as described in [34], is employed.

The output of the IEKF is an estimate of the current system state, i.e., 3-D structure and 3-D motion, as well as the uncertainty associated with this estimate in the form of a covariance matrix. Exploiting this information, we can formulate a statistical test that is able to decide whether or not each point X_o^i is consistent with the estimated dynamics model (3.55) (see [32]). Thus, we are able to identify points that do not belong to the currently tracked object.

This statistical test can be illustrated on a simple example. We have evaluated our method in an image sequence of the, Istituto Elettrotecnico Nazionale (IEN) Galileo Ferraris image database.[5] Figure 3.15 shows the first stereo image pair from the sequence, which contains a car driving in front of the observer at a nearly constant relative distance.

Six points on the preceding vehicle and two points on the road close to the vehicle have been manually selected. The statistical test to separate the selected points is shown in Figure 3.16. The two points on the road were detected after seven and eleven iterations, respectively. Additionally, after the exclusion of the two outliers, the Kalman filter provides the true relative velocity of the vehicle.

In this section different approaches to scene segmentation and object detection by interpreting motion data have been described. Simple scene segmentation can be obtained by evaluating the similarity of 3-D motion vectors; however, this approach will fail in many cases. General scenes can be handled by exploiting the epipolar geometry — either using essential matrices, which yield only a necessary condition for motion segmentation, or by using the trifocal constraint, which gives a sufficient condition but requires three images. We have also shown a simple example of how independent objects can be separated from stereo image streams. This approach is based on a simultaneous 3-D reconstruction and 3-D motion estimation. In each of the described methods, information about the uncertainty of the

---

[5] The database can be accessed via the World Wide Web at http://www.ien.it/is/is.html.

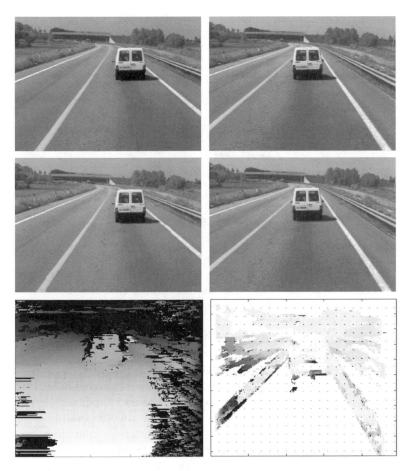

Figure 3.15 Sample data. Top: left and right stereo images at time t; Middle: left and right images at time $t + 1$; Bottom: extracted disparity image and 2-D motion field of the right camera.

2-D motion vectors can be incorporated in the segmentation process to obtain meaningful results.

3.5   Conclusion

Motion information plays a key role in numerous image-processing tasks. While image sequence encoding benefits from the strong bindings of intensity along motion trajectories, measurement applications employ motion as an observation that incorporates information about 3-D geometry and 3-D motion of the real world.

   2-D motion in image sequences is mainly induced from 3-D motion in the real world. Since rigid motion dominates nonrigid motion for most objects in our world by far, it is worthwhile to analyze motion of rigid objects

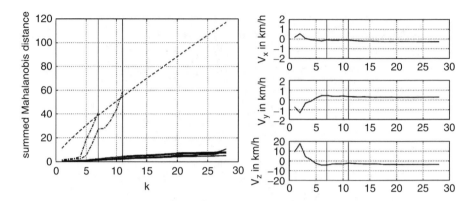

Figure 3.16   Statistical GOF test and estimated relative velocity. Two points that do not belong to the tracked vehicle are detected and removed from the estimation process.

of arbitrary 3-D shape. The epipolar constraint reduces motion vectors between two images to a single degree of freedom plus five parameters per rigid object that are comprised in the epipolar matrix. The trifocal tensor constrains image motion between multiple images even more strongly since it intrinsically encodes the exact position of observed features.

The computationally fastest class of approaches to motion computation is the family of gradient-based methods. These approximate image intensity locally by a low-order polynomial. Therefore, these methods may experience problems with large motion. On the other hand, they inherently offer sub-pixel accuracy for small motion amplitudes. Intensity-matching methods directly compare regions in different images according to some similarity measure and are robust against large motion amplitudes. They cannot cope, however, with strong spatial variations of motion. Such problems can be overcome by extraction and matching of features that are invariant with respect to the expected kind of variations. Shape context matching has been discussed as a method from this latter category.

Uncertainty assessment plays an important role in motion computation. From an extreme point of view, one can even argue that an estimate without accompanying uncertainty measure is absolutely useless. Motion covariance provides a simple yet efficient means to describe uncertainty. It can be integrated naturally into the motion computation process for all methods discussed.

Four concise applications for motion computation from fields that differ in nature have been discussed. Beyond large differences in requirements, it is argued that motion computation can be formulated in a unique mathematical framework. The individual applications take the benefit of uncertainty information in different ways. In image sequence coding, uncertainty quantifies the content of information that could be gathered by the motion computation process. Typically, uncertainty is large in poorly structured regions. Motion parameters that have little uncertainty should be prioritized for data rate assignment over motion vectors with large uncertainty. For motion field encoding with a segmentation-based transform, this leads to a weighted transform, in which the encoding error of each individual motion vector is weighted with its inverse covariance. In 3-D sensing, uncertainty is employed to detect gross outliers and to combine individual measurements. In tracking, uncertainty of measurements is propagated into uncertainty of object parameters. This uncertainty is then used to integrate novel observations with appropriate weight. Despite the differences in the task and usage of motion computation among the video encoding and video measurement applications, we see that both worlds benefit from merging their methods as, at last, the same physical quantity, motion, is sought.

References

[1]   E. Dubois and J. Konrad. Estimation of 2-D motion fields from image sequences with application to motion-compensated processing. In M.I. Sezan and R.L. Lagendijk, editors, *Motion Analysis and Image Sequence Processing*, chapter 3, pages 53–87. Kluwer Academic Publishers, Dordrecht, 1993.

[2]   T. Poggio and V. Torre. Ill-posed problems and regularization analysis in early vision. Technical Report 773, MIT Artificial Intelligence Laboratory, 1984.

[3]   R.I. Hartley. Lines and points in three views — a unified approach. *ARPA Image Understanding Workshop*, II:1009–1016, 1994.

[4]   R.I. Hartley and A. Zisserman. *Multiple View Geometry in Computer Vision*. Cambridge University Press, Cambridge, 2000.

[5]   H. Shariat and K.E. Price. Motion estimation with more than two frames. *IEEE Transactions on Pattern Analysis and Machine Intelligence*, 12(5):417–434, 1990.

[6]   H.-H. Nagel. On the estimation of optical flow: relations between different approaches and some new results. *Artificial Intelligence*, 33:299–324, 1987.

[7]   J.L. Barron, D.J. Fleet, and S. Beauchemin. Performance of optical flow techniques. *International Journal of Computer Vision*, 12(1):43–77, 1994.

[8]   C. Stiller and J. Konrad. On models, criteria, and search strategies for motion estimation in image sequences. *IEEE Signal Processing Magazine*, 7;9:70–91; 116–117, 1999.

[9]   B.K.P. Horn and B. G. Schunck. Determining optical flow. *Artificial Intelligence*, 17:185–203, 1981.

[10] S. Geman and D. Geman. Stochastic relaxation, Gibbs distributions, and the Bayesian restoration of images. *IEEE Transactions on Pattern Analysis and Machine Intelligence*, 6(6):721–741, 1984.

[11] J. Konrad and E. Dubois. Bayesian estimation of motion vector fields. *IEEE Transactions on Pattern Analysis and Machine Intelligence*, 14:910–927, 1992.

[12] B. Jähne. *Digital Image Processing: Concepts, Algorithms, and Scientific Applications*. Springer-Verlag, Berlin; Heidelberg, 5th edition, 2001.

[13] H. Haußecker and H. Spies. Motion. In B. Jähne, H. Haußecker, and P. Geißler, editors, *Handbook of Computer Vision and Applications*. Academic Press, New York, 1999.

[14] A. Singh. An estimation-theoretic framework for image-flow computation. *Proceedings of the IEEE International Conference on Computer Vision*, pages 168–177, 1990.

[15] M. Hötter, R. Mester, and M. Meyer. Detection of moving objects using a robust displacement estimation including a statistical error analysis. *International Conference on Pattern Recognition*, pages 249–255, 1996.

[16] S. Belongie and J. Malik. Matching with shape contexts. *IEEE Workshop on Content-Based Access of Image and Video Libraries*, 2000.

[17] A. Thayananthan et al. Shape context and chamfer matching in cluttered scenes. *Conference on Computer Vision and Pattern Recognition*, 2003.

[18] T. Sikora. MPEG digital video-coding standards. *IEEE Signal Processing Magazine*, 14(5):82–100, 1997.

[19] W.I. Grosky and R. Jain. A pyramid-based approach to segmentation applied to region matching. *IEEE Transactions on Pattern Analysis and Machine Intelligence*, 8:639, 1986.

[20] J. Ostermann. Modeling 3-D moving objects for an analysis-synthesis coder. *SPIE — Symposium on Electronic Imaging*, pages 1–10, 1990.

[21] E. Mémin, P. Pérez, and D. Machecourt. Dense estimation and object-oriented segmentation of the optical flow with robust techniques. Technical Report 991, Institut de Recherche en Informatique et Systèmes Aléatoires, March 1996.

[22] C. Stiller. Object oriented video coding employing dense motion fields. In *Proceedings of the IEEE International Conference on Acoustics, Speech and Signal Processing*, V:273–276, 1994.

[23] C.E. Shannon. A mathematical theory of communication. *The Bell System Technical Journal*, 27:379–423 & 623–656, Jul. & Oct. 1948.

[24] C. Stiller and D. Lappe. Gain/cost controlled displacement-estimation for image sequence coding. In *Proceedings of the IEEE International Conference on Acoustics Acoustics, Speech and Signal Proceeding*, IV:2729–2732, 1991.

[25] B. Girod. Rate-constrained motion estimation. *Proceedings of SPIE Visual Communications and Image Processing*, 2308:1026–1034, 1994.

[26] C. Stiller. Object-based estimation of dense motion fields. *IEEE Transactions on Image Processing*, 6(2):234–250, 1997.

[27] P.J. Huber. *Robust Statistics*. Wiley Series in Probability and Mathematical Statistics. John Wiley & Sons, New York, 1st edition, 1981.

[28] S. Kammel. Deflectometry for quality control of specular surfaces. *TM - Technisches Messen*, 2003.

[29] G. Farin. *Curves and Surfaces for Computer-Aided Geometric Design*. Academic Press, New York, 1997.

[30] P.H.S. Torr. *Outlier Detection and Motion Segmentation*. Ph.D. thesis, University of Oxford, 1995.

[31] O. Faugeras. *Three dimensional computer vision: a geometric viewpoint*. MIT Press, Cambridge, 1st edition, 1995.

[32] T. Dang, C. Hoffmann, and C. Stiller. Fusing optical flow and stereo disparity for object tracking. *IEEE International Conference on Intelligent Transportation Systems*, 2002.

[33] A. Chiuso and S. Soatto. Motion and structure from 2-D motion causally integrated over time: Analysis. *IEEE Transactions on Robotics and Automation*, 24:523–535, 2000.

[34] Y. Bar-Shalom and T. Fortmann. *Tracking and Data Association*. Academic Press, New York, 1988.

[35] A. Fusiello, E. Trucco, and A. Verri. A compact algorithm for rectification of stereo pairs. *Machine Vision and Applications*, 12(1):16–22, 2000.

chapter 4

Motion analysis and displacement estimation in the frequency domain

Luca Lucchese and Guido Maria Cortelazzo

Contents

4.1 Introduction

The frequency domain investigation of 2-D motion in image sequences has a long tradition in optical signal processing [1–6]. Among the advantages offered by this domain is the representation of translational displacements in terms of phase shifts. The frequency domain has later attracted the attention of the image-processing and computer vision communities, which exploited this property for purely translational motion with constant velocity in various applications such as television [7], neural modeling of motion perception [8], and velocity estimation [9]. Translational motion depending on more complicated temporal laws [10, 11], rotational motion [12, 13], and affine motion [14] have also been analyzed in the frequency domain.

The structural properties of the Fourier transforms of moving objects have been exploited to estimate motion parameters. The phase correlation

0-8493-1526-3/2004/$0.00+$1.50
© 2004 by CRC Press LLC

method was proposed to retrieve translational displacements from phase shifts [15]. Extensions of this method have been advanced to estimate rotational motion, as well [16, 17, 18]. These frequency domain methods have the important advantage that they use all the information available and not just the information associated with sets of features. Therefore, the delicate issues associated with the correspondence problem can be avoided and the methods can be rather robust against noise or image impairment, as the analysis of [19] and [20] highlights. Furthermore, if the algorithmic steps necessary to extract the motion parameters are able to preserve the original information, these techniques may be potentially highly accurate as well [20].

This chapter systematically presents fundamental results in planar motion analysis and estimation. Section 4.2 introduces the frequency domain analysis of planar rigid motion. Section 4.3 extends the analysis to the case of planar affine motion. Section 4.4 presents an algorithm for estimating planar roto-translational displacements, and Section 4.5 presents an algorithm for estimating affine displacements. Section 4.6 draws the conclusions and provides an evolutionary perspective on motion analysis and estimation in the frequency domain.

4.2   *Frequency domain analysis of planar rigid motion*

This section reviews the spatio-temporal model of [13] for analyzing the effects of 2-D rigid motion (both translational and rotational) in the frequency domain. The model is first derived for the case of 2-D rigid motion in which both translational and rotational components follow polynomial temporal laws with infinite duration and then is extended to the case of 2-D rigid motion of the same type applied for a finite amount of time. Figures 4.1(a) and (b) show two images, taken at different times, of a planar rigid object that rotates and translates on a uniform background on the plane indexed by the Cartesian reference system $x0y$; this system is fixed, i.e., it does not move over time. The center of rotation of the object at time t, denoted as $C(t)$, is assumed to translate on the plane. A second Cartesian coordinate reference system $x_0 C(t) y_0$ translates on the plane along with $C(t)$, maintaining its axes parallel to those of $x0y$. By denoting a given time instant as t_0, the object at $t = t_0$ can be described with respect to $x_0 C(t_0) y_0$ as a signal $\ell_s(x_0)$, $x_0 \doteq [x_0 \ y_0]^T \in \mathbb{R}^2$ (this notation stresses the time invariance of the signal $\ell_s(x_0)$). At a generic time instant $t > t_0$, the object can instead be represented, with respect to $x0y$, as a signal $\ell(x,t)$, $x \doteq [x \ y]^T \in \mathbb{R}^2, t \in \mathbb{R}$. The relationship between the coordinates x_0 and x of a point in the two reference systems (see Figure 4.1(b)) is therefore

$$x = R(t)x_0 + b(t), \tag{4.1}$$

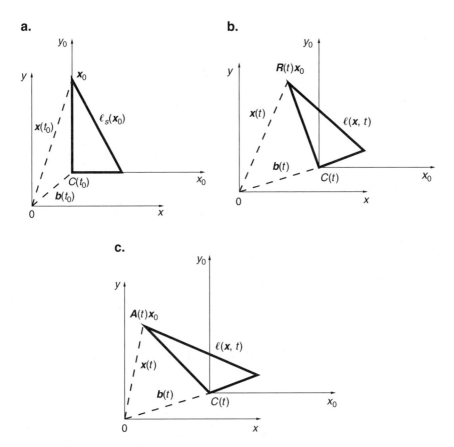

Figure 4.1 (a) Object at time $t = t_0$; (b) object at time $t > t_0$ during the rigid motion described by Equation (4.1); (c) object at time $t > t_0$ during the affine motion described by Equation (4.32).

where

$$R(t) = \begin{bmatrix} \cos\theta(t) & -\sin\theta(t) \\ \sin\theta(t) & \cos\theta(t) \end{bmatrix} \in \mathrm{SO}(2), \tag{4.2}$$

$$\theta(t) = \sum_{l=1}^{L} \omega_l \frac{t^l}{l!}, \quad \omega_l \in \mathbb{R}, \quad -\infty < t < +\infty, \tag{4.3}$$

$$b(t) = \sum_{m=1}^{M} v_m \frac{t^m}{m!}, \quad v_m \in \mathbb{R}^2, \quad -\infty < t < +\infty. \tag{4.4}$$

$R(t)$ is the rotation matrix at time t, $\theta(t)$ expresses the time dependence of the rotation angle, $b(t)$ is the position of the rotation center $C(t)$, and $\mathrm{SO}(2)$

denotes the group of the 2×2 special orthogonal matrices. The signals $\ell(x,t)$ and $\ell_s(x_0)$ thus relate as

$$\ell(x,t) = \ell_s(R^{-1}(t)(x - b(t))). \tag{4.5}$$

The 3-D Fourier transform (FT) of Equation (4.5) expressed in orthogonal Cartesian coordinates is

$$L(k,f) \doteq F[\ell(x,t)|k,f] = \int_{R^3} \ell(x,t)\exp\left(-j2\pi(k^T x + ft)\right) dx dt$$

$$= \int_{R^3} \ell_s(R^{-1}(t)(x - b(t)))\exp\left(-j2\pi(k^T x + ft)\right) dx dt, \tag{4.6}$$

where $k \doteq [k_x\ k_y]^T \in R^2$ is the vector of the spatial frequencies and $f \in R$ denotes the temporal frequency.

Since the signal $\ell_s(x_0)$ is angularly periodic with period 2π if expressed in polar coordinates, we consider its polar representation $\bar{\ell}_s(r, \varphi - \theta(t))$. By denoting the polar coordinates in the spatial domain with respect to the reference system $x_0 C(t_0) y_0$ as $x_0 = r\cos\varphi$, $y_0 = r\sin\varphi$, $r \geq 0$, $0 \leq \varphi < 2\pi$, and in the frequency domain as $k_x = k_r \cos k_\varphi$, $k_y = k_r \sin k_\varphi$, $k_r \geq 0$, $0 \leq k_\varphi < 2\pi$, the polar version of Equation (4.6) is

$$\bar{L}(k_r, k_\varphi, f) \doteq L(k_r \cos k_\varphi, k_r \sin k_\varphi, f)$$

$$= \int_{-\infty}^{+\infty} \int_{-\pi}^{\pi} \int_0^\infty r \bar{\ell}_s(r, \varphi - \theta(t)) \exp\left(-j2\pi r k_r \cos(\varphi - k_\varphi)\right) \tag{4.7}$$

$$\cdot \exp\left(-j2\pi k_r (x_C(t)\cos k_\varphi + y_C(t)\sin k_\varphi)\right) \exp\left(-j2\pi ft\right) dr\, d\varphi\, dt.$$

Since the function $\bar{\ell}_s(r, \varphi - \theta(t))$ has period 2π along its angular coordinate, it can be expressed in terms of its Fourier series as

$$\bar{\ell}_s(r, \varphi - \theta(t)) = \sum_{n=-\infty}^{+\infty} a_n(r)\exp\left(jn(\varphi - \theta(t))\right), \tag{4.8}$$

where

$$a_n(r) = \frac{1}{2\pi} \int_{-\pi}^{\pi} \bar{\ell}_s(r, \varphi)\exp\left(-jn\varphi\right) d\varphi. \tag{4.9}$$

This representation of $\bar{\ell}_s(r, \varphi - \theta(t))$ allows the separation of the translational and rotational components in Equation (4.7); in fact, by means of Equation (4.8), Equation (4.7) can be rewritten (see [13]) as

$$\bar{L}(k_r, k_\varphi, f) = \sum_{n=-\infty}^{+\infty} \bar{L}_{sn}(k_r, k_\varphi) \left\{ \left[\bigotimes_{m=1}^{M} \bar{T}_m(k_r, k_\varphi, .) \right] \otimes \left[\bigotimes_{l=1}^{L} \mathcal{R}_l(., n) \right] \right\}(f), \quad (4.10)$$

where

$$\bar{L}_{sn}(k_r, k_\varphi) \doteq 2\pi \exp\left(jn(k_\varphi - \pi/2)\right) \bar{S}_n(2\pi k_r), \quad (4.11)$$

$$\bar{S}_n(2\pi k_r) = \int_0^\infty r\, a_n(r)\, J_n(2\pi r k_r)\, dr, \quad (4.12)$$

$$J_n(2\pi r k_r) = \frac{1}{2\pi} \int_{-\pi}^{\pi} \exp\left(j(n\varphi - 2\pi r k_r \sin\varphi)\right) d\varphi, \quad (4.13)$$

$$\bar{T}_m(k_r, k_\varphi, f) \doteq \mathcal{F}\left[\exp\left(-j2\pi k_r \left[\cos k_\varphi \ \sin k_\varphi\right] v_m \frac{t^m}{m!} \right) | f \right], \quad (4.14)$$

$$\mathcal{R}_l(f, n) \doteq \mathcal{F}\left[\exp\left(-jn\omega_l \frac{t^l}{l!} \right) | f \right]. \quad (4.15)$$

The symbol $\otimes$ denotes the convolution operation with respect to f and $\otimes_{m=1}^{M}$ indicates the convolution with respect to f of M functions depending on this variable. It should be noted that the function $\bar{S}_n(2\pi k_r)$ is the Hankel transform of $a_n(r)$ computed at $2\pi k_r$ and $J_n(2\pi r k_r)$ is the nth-order Bessel function of the first kind evaluated at $2\pi r k_r$.

Equation (4.10) can be expressed in a Cartesian format as

$$L(k, f) = \sum_{n=\infty}^{+\infty} L_{sn}(k) \mathcal{P}^{(M,L)}(k, f, n), \quad (4.16)$$

where

$$L_{sn}(k) \doteq \bar{L}_{sn}\left(\sqrt{k_x^2 + k_y^2}, \arctan(k_y/k_x)\right), \quad (4.17)$$

$$\mathcal{P}^{(M,L)}(k, f, n) \doteq \left\{ \left[\bigotimes_{m=1}^{M} T_m(k, .) \right] \otimes \left[\bigotimes_{l=1}^{L} \mathcal{R}_l(., n) \right] \right\}(f), \quad (4.18)$$

and

$$T_m(\mathbf{k},f) \doteq \overline{T}_m\left(\sqrt{k_x^2 + k_y^2}, \arctan(k_y / k_x), f\right). \tag{4.19}$$

Equations (4.10) and (4.16) show that the Fourier transform of a planar object undergoing a planar rigid motion of infinite duration on a uniform background consists of a series expansion in which the contribution of the object's shape, encoded by the terms $L_{sn}(\mathbf{k})$ in Equation (4.16), is separated from the contribution of the object's motion, encoded by the terms $\mathcal{P}^{(M,L)}(\mathbf{k},f,n)$ in Equation (4.16). It should also be noted that the terms encoding shape information depend on the spatial frequencies, whereas the terms encoding motion information depend on spatial as well as on temporal frequencies. In particular, the M translational terms (expressed by Equation (4.19) in the case of Equation (4.16)) are convolved among themselves, and the result is convolved with the convolution of the L rotational terms (expressed by Equation (4.15)). The rotational terms, unlike the translational components, have a direct dependence on the index n of the series expansion.

As an example, we derive the expression of the Fourier transform of an object undergoing uniform translational motion. For $m = 1$, Equation (4.19) through Equation (4.14) yields

$$T_1(\mathbf{k},f) = \delta(f + \mathbf{k}^T \mathbf{v}_1) \tag{4.20}$$

and for $l = 1$, Equation (4.15) gives

$$\mathcal{R}_1(f,n) = \delta(f + n\omega_1 / 2\pi), \tag{4.21}$$

where $\delta(.)$ denotes Dirac's impulse function. Therefore, Equation (4.18) reduces to

$$\mathcal{P}^{(1,1)}(\mathbf{k},f,n) \doteq [T_1(\mathbf{k},.) \otimes \mathcal{R}_1(.,n)](f) = \delta(f + \mathbf{k}^T \mathbf{v}_1 + n\omega_1 / 2\pi) \tag{4.22}$$

and Equation (4.16) becomes

$$L(\mathbf{k},f) = \sum_{n=-\infty}^{+\infty} L_{sn}(\mathbf{k}) \delta(f + \mathbf{k}^T \mathbf{v}_1 + n\omega_1 / 2\pi) \tag{4.23}$$

Equation (4.23) shows that the components $L_{sn}(\mathbf{k})$ of the Fourier transform of the signal $\ell_s(\mathbf{x}_0)$ are projected onto the bundle of parallel planes[1]

$$f + \mathbf{k}^T \mathbf{v}_1 + n\omega_1 / 2\pi = 0, \quad n \in \mathbb{Z}. \tag{4.24}$$

---

[1] Symbol $\mathbb{Z}$ denotes the set of integers.

The slant of these planes depends only on the translational velocity v_1 and their spacing[2]

$$d^{(1)} = \frac{|\omega_1|}{2\pi\sqrt{\|v_1\|^2 + 1}} \tag{4.25}$$

is determined by both v_1 and ω_1.

In Equation (4.23), it is worth observing that if $L_s(k)$ is spatially band limited, the presence of the velocity v_1 leads to a transform $L(k,f)$, which is band limited with respect to the temporal frequency only if the series of Equation (4.8) has a finite number of terms. This is because velocities are assumed to be applied for an infinite amount of time, which is a nonphysical situation.

If the object of Figure 4.1(a) is subject to a rigid motion for a finite amount of time, e.g., over $[0,T]$, the right-hand side of Equation (4.5) has to be multiplied by the temporal window

$$w(t) \doteq \left(\frac{t - T/2}{T} \right) = \begin{cases} 1, & if\ 0 \le t \le T, \\ 0, & elsewhere. \end{cases} \tag{4.26}$$

By denoting the Fourier transform of $w(t)$ as

$$W(f) \doteq \mathcal{F}\big[w(t)|f\big] = T\exp(-j\pi fT)\operatorname{sinc}(fT), \tag{4.27}$$

in the case of finite temporal duration, the expression corresponding to Equation (4.16) becomes

$$L(k,f) \doteq \mathcal{F}\big[\ell(x,t)w(t)|k,f\big] = \sum_{n=\infty}^{+\infty} L_{sn}(k)\mathcal{P}_W^{(M,L)}(k,f,n), \tag{4.28}$$

with

$$\mathcal{P}_W^{(M,L)}(k,f,n) \doteq \big(\mathcal{P}^{(M,L)}(k,.,n) \otimes W(.)\big)(f). \tag{4.29}$$

For $M = L = 1$, Equation (4.29) yields

[2] Symbol $\|\ \|$ denotes the Euclidean norm of vectors.

$$\mathcal{P}_W^{(1,1)}(\boldsymbol{k},f,n) = \left(\mathcal{P}^{(1,1)}(\boldsymbol{k},.,n) \otimes \mathcal{W}(.)\right)(f)$$

$$= \left(\delta(.+\boldsymbol{k}^T\boldsymbol{v}_1 + n w_1/2\pi) \otimes \mathcal{W}(.)\right)(f) = \mathcal{W}\left(f + \boldsymbol{k}^T\boldsymbol{v}_1 + n w_1/2\pi\right).$$

$$(4.30)$$

Therefore, the counterpart of Equation (4.23) is

$$L(\boldsymbol{k},f) = \sum_{n=-\infty}^{+\infty} L_{sn}(\boldsymbol{k}) \mathcal{W}(f + \boldsymbol{k}^T\boldsymbol{v}_1 + n\omega_1/2\pi). \qquad (4.31)$$

The impulse functions of Equation (4.23) are replaced in Equation (4.31) by *sinc* functions with the same argument, whose magnitudes attain their largest values on the planes defined by Equation (4.24).

The frequency domain analysis of 2-D rigid motion with infinite duration described by more complex polynomial laws requires the computation of the Fourier transforms of Equations (2.14) and (2.15) for $m \geq 2$ and $l \geq 2$ and of their convolution in Equation (4.18). The closed-form expressions of the Fourier transforms in Equation (4.14) can be found in [10] (for $m = 2,3$) and in [11] (for any m). The closed-form expressions of the Fourier transforms in Equation (4.15) can be obtained through simple algebraic substitutions from these results, since the two sets of equations share a similar structure. The analysis of 2-D rigid motion with infinite duration and constant velocity and/or acceleration is instead presented in [13].

4.3 Frequency domain analysis of planar affine motion

Affine deformations are a traditional tool for modeling nonrigid displacements, both local and global, between pairs of images. As such, they are used in a variety of applications, including video coding, shape from texture, and pattern recognition. This section presents an analytical tool for the frequency domain analysis of 2-D affine deformations of dynamic sequences. Figures 4.1(a) and (c) show two images, taken at different times, of an object that undergoes a planar affine transformation on a uniform background. For convenience, we use the same coordinate reference systems $x0y$ and $x_0 C(t) y_0$ defined in Section 4.2, and we directly consider the case of affine motion with a finite duration $[0,T]$. At time $t = t_0 = 0$, the object can be described with respect to $x_0 C(t_0) y_0$ as a signal $\ell_s(\boldsymbol{x}_0)$.

At a generic time instant $t_0 < t \leq T$ the object can be represented with respect to $x0y$ as a signal $\ell(\boldsymbol{x},t)$. The relationship between the coordinates $\boldsymbol{x}$ and $\boldsymbol{x}_0$ of a generic point in the two reference systems (see Figure 4.1(c)) is

$$\boldsymbol{x} = A(t)\boldsymbol{x}_0 + \boldsymbol{b}(t), \qquad 0 \leq t \leq T, \qquad (4.32)$$

where

$$A(t) = \begin{bmatrix} a_{11}(t) & a_{12}(t) \\ a_{21}(t) & a_{22}(t) \end{bmatrix}, \quad \det(A(t)) \neq 0,$$

(4.33)

$$a_{ij}(t) : \mathbb{R} \to \mathbb{R}, \quad a_{ii}(0) = 1 \quad a_{ij}(0) = 0 \ i \neq j,$$

and

$$b(t) = \begin{bmatrix} b_1(t) \\ b_2(t) \end{bmatrix}, \quad b_i(t) : \mathbb{R} \to \mathbb{R}, \quad b_i(0) = 0.$$

(4.34)

It should be observed that the constraints on the temporal functions $a_{ij}(t)$ and $b_i(t)$ in Equations (4.33) and (4.34) enforce the coincidence of x with x_0 at $t = t_0 = 0$. The two signals $\ell(x, t)$ and $\ell_s(x_0)$ relate as

$$\ell(x, t) = \ell_s(A^{-1}(t)(x - b(t)))w(t),$$

(4.35)

where $w(t)$ is the temporal window defined in Equation (4.26).

The 3-D Fourier transform of Equation (4.35), similar to Equation (4.6), is

$$L(k, f) \doteq \mathcal{F}\left[\ell(x, t) \mid k, f\right] =$$

$$= \int_{\mathbb{R}^3} \ell_s(A^{-1}(t)(x - b(t)))w(t)\exp\left(-j2\pi\left(k^T x + ft\right)\right) dx dt = \quad (4.36)$$

$$= \int_{-\infty}^{+\infty} \left|\det(A(t))\right| L_s(A(t)^T k)w(t)\exp\left(-j2\pi\left(k^T b(t) + ft\right)\right) dt,$$

where

$$L_s(k) \doteq \mathcal{F}\left[\ell_s(x) \mid k\right] = \int_{\mathbb{R}^2} \ell_s(x)\exp(-j2\pi k^T x) dx.$$

(4.37)

The integration with respect to the spatial variables is a result of the affine theorem for two-dimensional transforms [21]. The integration with respect to time for the general case cannot be further pursued because of the presence in Equation (4.36) of the two implicit functions of time $|\det(A(t))|$ and $L_s(A(t)^T k)$.

In order to solve the integration with respect to time, we adopt a linear time-dependence model for the affine deformation parameters $A(t)$ and $b(t)$. This simplistic representation may reasonably approximate general motion only for a rather short duration T. For large values of T, one can resort to a piecewise representation of the motion during $[0, T]$ (an approach similar to the one presented in [10] for 2-D translational motion with cubic temporal laws) and, thanks to the linearity of the Fourier transform, the affine motion can be analyzed over arbitrary time intervals by summing up the contributions of subintervals of [0,*T*]. This approach can also handle discontinuities and/or juxtapositions of totally different affine deformations.

The piecewise representation of motion may be readily formalized as follows; let us divide [0,*T*] into *I* intervals [t_i, t_{i+1}], $i = 0, \ldots, I-1$, with $t_0 \doteq 0$ and $t_I \doteq T$, and let us suppose that during the *i*th time interval, of duration $\Delta t_i \doteq t_{i+1} - t_i$, the motion may be represented by the model of Equation (4.35) with $A(t) = A_i(t)$ and $b(t) = b_i(t)$. In this case, Equation (4.35) can be rewritten as

$$\ell(x,t) = \sum_{i=0}^{I-1} \ell_s(A_i^{-1}(t)(x - b_i(t)))w_i(t), \qquad (4.38)$$

with

$$w_i(t) \doteq \mathrm{rect}\left(\frac{t - (t_i + t_{i+1})/2}{\Delta t_i}\right) = \begin{cases} 1, & \text{if } t_i \le t \le t_{i+1}, \\ 0, & \text{elsewhere,} \end{cases} \quad i = 1, \ldots, I-1. \quad (4.39)$$

The Fourier transform of Equation (4.38) is thus a simple generalization of Equation (4.36), i.e.,

$$L(k,f) = \sum_{i=0}^{I-1} \int_{-\infty}^{+\infty} |\det(A_i(t))| L_s(A_i(t)^T k) w_i(t) \exp(-j2\pi(k^T b_i(t) + ft)) dt. \quad (4.40)$$

A motion without discontinuities may be obtained by enforcing the two constraints $A_{i+1}(t_{i+1}) = A_i(t_{i+1})$ and $b_{i+1}(t_{i+1}) = b_i(t_{i+1})$, $i = 0, \ldots, I-2$.

For simplicity of analysis, and without any loss of generality, we will restrict ourselves to the consideration of the motion only during a single time interval as modeled by Equations (4.35) and (4.36).

An affine deformation with linear time dependence over [0,*T*] can be expressed as

$$A(t) = I + A_1 t \doteq \begin{bmatrix} 1 & 0 \\ 0 & 1 \end{bmatrix} + \begin{bmatrix} \alpha_{11} & \alpha_{12} \\ \alpha_{21} & \alpha_{22} \end{bmatrix} t, \quad \alpha_{ij} \in \mathbb{R}, \qquad (4.41)$$

and

$$b(t) = b_1 t \doteq \begin{bmatrix} b_1 \\ b_2 \end{bmatrix} t, \quad b_i \in \mathbb{R}. \qquad (4.42)$$

From Equation (4.41), one obtains

$$\det(A(t)) = 1 + \text{trace}(A_1)t + \det(A_1)t^2 \doteq d_0 + d_1 t + d_2 t^2. \tag{4.43}$$

The two constraints $\det(A_1) > 0$ and $\text{trace}^2(A_1) < 4\det(A_1)$ guarantee that $\det(A(t)) > 0 \ \forall t$, i.e., that the time-dependent affine deformation expressed by matrix $A(t)$ is not subject to degeneracies; this also implies that the object does not undergo any reflection while moving. Therefore, it is

$$|\det(A(t))| = \det(A(t)) \tag{4.44}$$

and Equation (4.43) can be used in Equation (4.36) as an explicit time expression for $|\det(A(t))|$.

Under the assumption that the function $L_s(A(t)^T k)$ has derivatives of any order with respect to time over $[0,T]$, one can separate the time variable t from the spatial frequencies k, which are embedded together as the argument of $L_s(.)$, by representing $L_s(A(t)^T k)$ through its McLaurin series with respect to time as

$$L_s(A(t)^T k) = \sum_{n=0}^{\infty} \sigma_n(k) \frac{t^n}{n!} \quad \text{with} \quad \sigma_n(k) \doteq \left(\frac{\partial^n L_s(A(t)^T k)}{\partial t^n} \right)_{t=0}. \tag{4.45}$$

A useful expression for the coefficients $\sigma_n(k)$ can be obtained as follows. For convenience, we call

$$\begin{bmatrix} u(k,t) \\ v(k,t) \end{bmatrix} \doteq A(t)^T k \tag{4.46}$$

and observe that at $t = 0$ from Equation (4.41), Equation (4.46) becomes

$$\begin{bmatrix} u(k,0) \\ v(k,0) \end{bmatrix} = A^T(0)k = k. \tag{4.47}$$

Based on the linear time-dependence model of Equation (4.41), the first partial derivatives of $u(k,t)$ and $v(k,t)$, with respect to time, are

$$\begin{bmatrix} \dot{u}(k) \\ \dot{v}(k) \end{bmatrix} \doteq \frac{\partial}{\partial t} \begin{bmatrix} u(k,t) \\ v(k,t) \end{bmatrix} = A_1^T k \tag{4.48}$$

whereas their higher-order partial derivatives are all equal to zero, i.e.,

$$\frac{\partial^q}{\partial t^q} \begin{bmatrix} u(k,t) \\ v(k,t) \end{bmatrix} = 0, \quad q \geq 2. \tag{4.49}$$

The first partial derivative of $L_s(.)$ with respect to time can thus be written as

$$\frac{\partial L_s(A(t)^T k)}{\partial t} = \frac{\partial L_s(u,v)}{\partial u}\frac{\partial u}{\partial t} + \frac{\partial L_s(u,v)}{\partial v}\frac{\partial v}{\partial t}, \tag{4.50}$$

its second partial derivative as

$$\frac{\partial^2 L_s(A(t)^T k)}{\partial t^2} = \frac{\partial^2 L_s(u,v)}{\partial u^2}\left(\frac{\partial u}{\partial t}\right)^2 + 2\frac{\partial^2 L_s(u,v)}{\partial u \partial v}\frac{\partial u}{\partial t}\frac{\partial v}{\partial t} + \frac{\partial^2 L_s(u,v)}{\partial v^2}\left(\frac{\partial v}{\partial t}\right)^2$$

(4.51)

and, in general, its nth partial derivative as

$$\frac{\partial^n L_s(A(t)^T k)}{\partial t^n} = \sum_{p=0}^{n}\binom{n}{p}\frac{\partial^{n-p} L_s(u,v)}{\partial u^{n-p}}\left(\frac{\partial u}{\partial t}\right)^{n-p}\frac{\partial^p L_s(u,v)}{\partial v^p}\left(\frac{\partial v}{\partial t}\right)^p, \tag{4.52}$$

where $\binom{n}{p} = \dfrac{n!}{p!(n-p)!}$ are the binomial distribution coefficients. Therefore, the coefficients $\sigma_n(k)$ of the series in Equation (4.45) can be expressed as

$$\sigma_n(k) = \left(\frac{\partial^n L_s(A(t)^T k)}{\partial t^n}\right)_{t=0} = \sum_{p=0}^{n}\binom{n}{p}\frac{\partial^{n-p} L_s(k)}{\partial k_x^{n-p}}\dot{u}(k)^{n-p}\frac{\partial^p L_s(k)}{\partial k_y^p}\dot{v}(k)^p =$$

$$= (-j2\pi)^n \sum_{p=0}^{n}\binom{n}{p}M_x^{(n-p)}(k)M_y^{(p)}(k)\dot{u}(k)^{n-p}\dot{v}(k)^p, \tag{4.53}$$

where

$$M_x^{(n-p)}(k) \doteq \mathcal{F}\left[x_0^{n-p}\ell_s(x_0)|k\right] \text{ and } M_y^{(p)}(k) \doteq \mathcal{F}\left[y_0^p\ell_s(x_0)|k\right]. \tag{4.54}$$

From the explicit expressions of $|\det(A(t))|$ and $L_s(A(t)^T k)$ with respect to time, the integral in Equation (4.36) yields

$$L(k,f) = \sum_{n=0}^{\infty}\frac{\sigma_n(k)}{n!}\sum_{m=0}^{2}d_m\int_{-\infty}^{+\infty}t^{n+m}w(t)\exp\left(-j2\pi\left(f+k^T b_1\right)t\right)dt =$$

(4.55)

$$= \sum_{n=0}^{\infty}\sum_{m=0}^{2}d_m\frac{\sigma_n(k)}{n!}\frac{1}{(-j2\pi)^{n+m}}\left(\frac{\partial^{n+m}W(v)}{\partial v^{n+m}}\right)_{v=f+k^T b_1},$$

where $\mathcal{W}(.)$ is defined by Equation (4.27). The derivatives[3] of this function can easily be computed, allowing $L(k,f)$ to be expressed in a closed form. In principle, an infinite number of terms have to be summed up to obtain this transform. However, the contribution of the functions $\sigma_n(k)$ in Equation (4.55) is expected to decrease as n increases so that a limited set of terms has to be computed. It is also important to observe that the derivatives of the *sinc* function $\mathcal{W}(v)$ in Equation (4.55) are smooth functions whose amplitude decreases as their order increases. This provides an additional justification for the use of a limited number of terms in the series expansion of Equation (4.55).

Another important remark concerns the common structural properties of the signals in Equations (4.31) and (4.55). In both cases, the translational component of motion determines the tilt of the spectra, which taper in a direction orthogonal to the plane $f + k^T v_1 = 0$ for rigid motion and to the plane $f + k^T b_1 = 0$ for affine motion. Additional comments on the structure of the Fourier transform in Equation (4.55) can be found in [14], where these results were originally presented.

4.4 Frequency domain estimation of planar roto-translational displacements

The frequency domain estimation of 2-D rigid motion offers two important advantages that have been recognized for a long time [16, 17, 18]: robustness to noise and decoupling of the rotational and translational components. The latter property is intrinsic in the structure of the Fourier representation of signals [22] and allows the estimation of rotation from the FT magnitude of the object undergoing rigid motion.

This section highlights the rationale behind the frequency domain algorithm advanced in [20] for estimating 2-D rigid motion. The details are omitted here for space constraints, and the interested reader is referred to the original paper for an in-depth analysis of the algorithm. The method is based upon the following property: given two images of an object relatively rotated and translated, the difference between the FT magnitude of one image and the mirror version, with respect to either frequency axis, of the FT magnitude of the other shows a pair of orthogonal zero-crossing lines. These lines are rotated with respect to the frequency axes by an angle that is half the rotation angle. Therefore, the estimation of planar rotation turns out to be equivalent to the detection of these two zero-crossing lines.

The recovery of the rotation angle is accomplished by a three-stage coarse-to-fine procedure, which achieves an accuracy of the order of a few hundredths of a degree in the absence of noise. A technique of phase correlation type [15] is subsequently adopted in order to discriminate between pairs of possible solutions for the rotation angle and to estimate the translational displacement. The analysis reported in [20] reveals that the two

---

[3] The zeroth derivative of $\mathcal{W}(.)$ is the function itself.

zero-crossing lines are very robust to noise at low frequencies. This enables the algorithm to yield estimates of the rotation angle with errors within a few tenths of a degree even with extremely low signal-to-noise ratios (SNRs). A similar robustness is achieved also against differences between the two images due to occlusions. Another important characteristic of this method is that it avoids signal conversions into polar coordinate systems.

Let $\ell_1(x)$ and $\ell_2(x)$, $x \in \mathbb{R}^2$, denote two scalar images related through a 2-D rigid motion as

$$\ell_2(x) = \ell_1(R(\theta)^{-1}x - b),\tag{4.56}$$

where

$$R(\theta) = \begin{bmatrix} \cos\theta & -\sin\theta \\ \sin\theta & \cos\theta \end{bmatrix} \in SO(2) \quad \text{and} \quad b \in \mathbb{R}^2.\tag{4.57}$$

As an example, Figures 4.2(a) and (c) show a pair of images $\ell_1(x)$ and $\ell_2(x)$ related through Equations (4.56) and (4.57) with $\theta = 25°$ and $b = 0$. The two images have size 301×301 pixels; the x-axis coincides with the central row and the y-axis (upward) coincides with the central column.

By denoting the 2-D Fourier transform of image $\ell_n(x)$ as

$$L_n(k) \doteq \mathcal{F}\big[\ell_n(x)\,|\,k\big] = \int_{\mathbb{R}^2} \ell_n(x)e^{-j2\pi k^T x}\,dx, \quad k \in \mathbb{R}^2, \quad n = 1,2,\tag{4.58}$$

Equation (4.56) in the frequency domain becomes [21]

$$L_2(k) = L_1(R(\theta)^{-1}k)e^{-j2\pi k^T R(q)^b},\tag{4.59}$$

and, by considering only the magnitudes,

$$\big|L_2(k)\big| = \big|L_1(R(\theta)^{-1}k)\big|\tag{4.60}$$

It can be noticed from Equation (4.60) that the two FT magnitudes are relatively rotated by the same angle θ as the two images. This can be graphically appreciated in the plots of Figures 4.2(b) and (d), which display the squared Fourier transform magnitudes $|L_1(k)|^2$ and $|L_2(k)|^2$ of the two images $\ell_1(x)$ and $\ell_2(x)$ of Figures 4.2(a) and (c), respectively.

If the two images differ not only by a 2-D rigid motion but also by an isotropic scaling, Equation (4.61) has to be replaced by

$$\ell_2(x) = \ell_1(\rho R(\theta)^{-1}x - b),\tag{4.61}$$

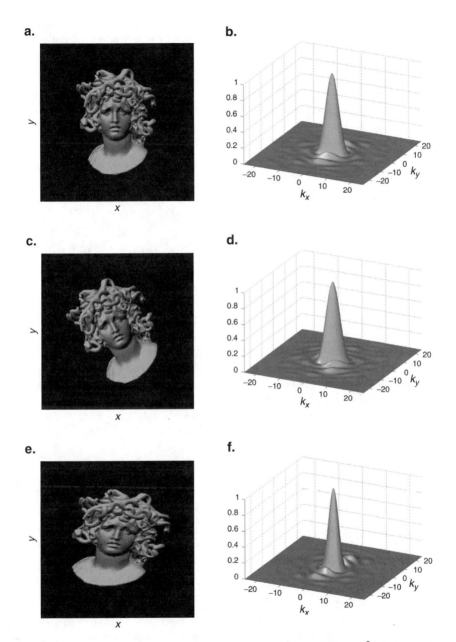

Figure 4.2 (a) Original image $\ell_1(x)$; (b) squared FT magnitude $\left|L_1(k)\right|^2$ relative to the image of (a); (c) image $\ell_2(x)$ obtained from $\ell_1(x)$ through the rigid motion of Equations (4.56) and (4.57) with $\theta = 25°$ and $b = 0$; (d) squared FT magnitude $\left|L_2(k)\right|^2$ relative to the image of (c); (e) image $\ell_2(x)$ obtained from $\ell_1(x)$ through the affine transformation of Equations (4.73) and (4.74) with $\mathbf{A} = \begin{bmatrix} 1.2 & 0.3 \\ -0.2 & 0.9 \end{bmatrix}$ and $b = 0$; (f) squared FT magnitude $\left|L_2(k)\right|^2$ relative to the image of (e).

where $\rho > 0$ is a scaling factor common to both coordinates. By using the scaling property of the Fourier transform [22], it is straightforward to compute the frequency domain counterpart of Equation (4.61) as

$$L_2(k) = \frac{1}{\rho^2} L_1(R(\theta)^{-1}(k/\rho)) e^{-j2\pi(k/\rho)^T R(\theta)^b},$$

(4.62)

that for magnitudes reads

$$|L_2(k)| = \frac{1}{\rho^2} |L_1(R(\theta)^{-1}(k/\rho))|.$$

(4.63)

From Equation (4.63), the scaling factor ρ can be estimated as the quotient $\hat{\rho} = \sqrt{L_1(0)/L_2(0)}$ of the two magnitudes for $k = 0$. By defining $L_1'(k) \doteq \frac{1}{\rho^2} L_1(k/\rho)$, Equation (4.63) can then be rewritten as

$$|L_2(k)| = |L_1'(R(\theta)^{-1}k)|.$$

(4.64)

Therefore after compensating for scaling, Equation (4.63), through Equation (4.64), reduces to Equation (4.60), and without any loss of generality we will refer only to Equation (4.60).

The algorithmic idea of [20] is based on a simple 1-D analogy for estimating translational displacements. This idea is extended to the estimation of planar rotations by defining the difference signal

$$\Delta(k) \doteq \frac{|L_1(k)|^2}{L_1^2(0)} - \frac{|L_2(Hk)|^2}{L_2^2(0)} = \frac{|L_1(k)|^2}{L_1^2(0)} - \frac{|L_1(R(\theta)^{-1}Hk)|^2}{L_1^2(0)},$$

(4.65)

where

$$H = \begin{bmatrix} \pm 1 & 0 \\ 0 & \mp 1 \end{bmatrix},$$

(4.66)

and the relationship in Equation (4.60) has been used. Matrix H performs the mirror reflection of the second transform with respect to the k_x-axis or the k_y-axis according to the pair of signs chosen for the entries of its principal diagonal.

The algorithm of [20] rests on a theorem about an important structural property of the function in Equation (4.65). Such a theorem is reported next along with its proof.

Theorem. Let $L(k)$, $k \in \mathbb{R}^2$, be a Fourier transform; if $L(k)$ has only Hermitian symmetry, then the function

$$\Delta(k) = \frac{|L(k)|^2}{L^2(0)} - \frac{|L(R(\theta)^{-1}Hk)|^2}{L^2(0)}, \tag{4.67}$$

where $R(\theta) \in SO(2)$ and H is the reflection matrix of Equation (4.66), is such that the locus $\Delta(k) = 0$ includes, as very distinctive features, the two orthogonal lines

$$\frac{k_y}{k_x} = -\tan\frac{\theta}{2} \quad and \quad \frac{k_y}{k_x} = 1/\tan\frac{\theta}{2}. \tag{4.68}$$

Proof. The function $L(k)$ has only Hermitian symmetry, i.e., $L(k) = L^*(-k)$. It follows that $\Delta(k) = 0$ if $R(\theta)^{-1}Hk = \pm k$. Therefore, the determination of the locus $\Delta(k) = 0$ turns out to be an eigenvalue problem. We can readily see that

$$R(\theta)^{-1}H = R(\theta)^T H = \begin{bmatrix} \pm\cos\theta & \mp\sin\theta \\ \mp\sin\theta & \mp\cos\theta \end{bmatrix}, \tag{4.69}$$

whose eigenvalues are $\lambda = +1$ and $\lambda = -1$; their eigenvectors are (in accordance with the pairs of signs)

$$v(\lambda = \pm1) = \begin{bmatrix} 1 \\ \frac{\cos\theta - 1}{\sin\theta} \end{bmatrix} \quad and \quad v(\lambda = \mp1) = \begin{bmatrix} 1 \\ \frac{\cos\theta + 1}{\sin\theta} \end{bmatrix}. \tag{4.70}$$

The eigenspace relative to $v(\lambda = \pm1)$ is the line through the origin with slope

$$\frac{k_y}{k_x} = \frac{\cos\theta - 1}{\sin\theta} = -\tan\frac{\theta}{2} \tag{4.71}$$

while the eigenspace relative to $v(\lambda = \mp1)$ is the line through the origin with slope

$$\frac{k_y}{k_x} = \frac{\cos\theta + 1}{\sin\theta} = 1/\tan\frac{\theta}{2}. \tag{4.72}$$

Thus, the locus $\Delta(k) = 0$ includes the two zero-crossing lines of Equation (4.68).

Figure 4.3(a) shows the plot of $z = \Delta(k)$ relative to the images of Figures 4.2(a) and (c); Figure 4.3(b) displays the corresponding locus $\Delta(k) = 0$. Both plots clearly show the two orthogonal lines of Equations (4.68); from the slopes of these lines, the algorithm of [20] returned $\theta = 25.02°$ as the estimate for θ. Once the rotational component of the rigid motion has been estimated, the translational component can be obtained by de-rotating either image and applying the phase correlation method [15]. In the example of this section, $\hat{b} = 0$ was the estimate found for b.

a.

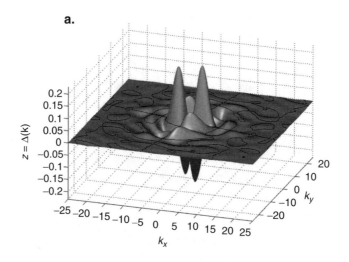

b.

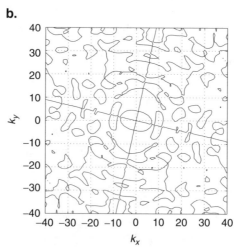

Figure 4.3 (a) Plot of $z = \Delta(k)$; (b) locus $\Delta(k) = 0$.

The literature offers other frequency domain methods for estimating 2-D rotations, which are comprehensively reviewed in [20].

4.5 Frequency domain estimation of planar affine displacements

Like the analysis of rigid motion presented in the previous section, the frequency domain analysis of affine transformations offers two main advantages: (1) it enables the estimation of the affine matrix to be decoupled from the estimate of the translational displacement [22]; and (2) it permits more robustness over feature-based methods because it is based on a low-level representation that uses all the image information [23, 24, 25, 26]. In this section, we present the frequency domain algorithm of [27] for estimating the global affine transformation between two images, which consists of two steps. In the first step, the estimation of the affine matrix is formulated as a nonlinear minimization problem. A fundamental stretching relationship between the radial projections of the image energies — defined as the squared Fourier transform magnitudes after a convenient normalization — lends itself to be interpreted as a curve-fitting problem in polar coordinates. In the second step, after compensating for the affine matrix in one of the two images, the translation vector can be recovered by a standard phase correlation technique.

Let $\ell_1(x)$ and $\ell_2(x)$, $x \in \mathbb{R}^2$, denote two scalar images related through a 2-D affine transformation[4] as

$$\ell_2(x) = \ell_1(A^{-1}x - b), \tag{4.73}$$

where

$$A = \begin{bmatrix} a_{11} & a_{12} \\ a_{21} & a_{22} \end{bmatrix} \in \mathbb{R}^{2\times2}, \quad \det(A) > 0, \quad b = \begin{bmatrix} b_1 \\ b_2 \end{bmatrix} \in \mathbb{R}^2. \tag{4.74}$$

As an example, Figure 4.2(e) shows an image $\ell_2(x)$ related to image $\ell_1(x)$ of Figure 4.2(a) through Equations (4.73) and (4.74) with parameters $A = \begin{bmatrix} 1.2 & 0.3 \\ -0.2 & 0.9 \end{bmatrix}$ and $b = 0$.

By applying the 2-D Fourier transform defined in Equation (4.58), Equation (4.73) in the frequency domain becomes [21]

$$L_2(k) = \det(A)L_1(A^T k)e^{-j2\pi k^T Ab}. \tag{4.75}$$

---

[4] The constraint det (**A**) > 0, like the constraint det **A**(t) > 0 in Section 4.3, preserves the relative orientation of the two images.

Similar to Equation (4.59) for 2-D rigid motion, Equation (4.75) shows that matrix A can be estimated independently of b by considering only the magnitudes of the transforms in Equation (4.75), i.e.,

$$\left|L_2(k)\right| = \det(A)\left|L_1(A^T k)\right|. \tag{4.76}$$

It should be noticed that $D \doteq \det(A)$ can be estimated as $\hat{D} \doteq L_2(0)/L_1(0)$, i.e., by computing the quotient of the two transforms for $k = 0$.

By squaring Equation (4.76) and by denoting with $E_n(k) \doteq \left|L_n(k)\right|^2$, $n = 1,2$, the energy spectra of the two images [22], it is $\mathcal{E}_2(k) = \det(A)\mathcal{E}_1(A^T k)$ where $\mathcal{E}_1(k) \doteq E_1(k)/E_1(0)$ and $\mathcal{E}_2(k) \doteq \det(A)E_2(k)/E_2(0)$ are the two energies conveniently normalized. In the polar coordinate system $\rho = \sqrt{k_x^2 + k_y^2}$, $\theta = \arctan\left(k_y/k_x\right)$, the relationship between the normalized energies becomes

$$\bar{\mathcal{E}}_2(\rho,\theta) = \det(A)\bar{\mathcal{E}}_1\big(\lambda(\theta;A)\rho,\varphi(\theta;A)\big), \tag{4.77}$$

with $\bar{\mathcal{E}}_n(\rho,\theta) \doteq \mathcal{E}_n(\rho\cos\theta, \ \rho\sin\theta)$, $n = 1,2$,

$$\lambda(\theta;A) \doteq \left(\big(a_{11}\cos\theta + a_{21}\sin\theta\big)^2 + \big(a_{12}\cos\theta + a_{22}\sin\theta\big)^2\right)^{1/2}, \tag{4.78}$$

and

$$\varphi(\theta;A) \doteq \arctan\left(\frac{a_{12}\cos\theta + a_{22}\sin\theta}{a_{11}\cos\theta + a_{21}\sin\theta}\right). \tag{4.79}$$

Function $\lambda(\theta;A)$ determines the relative radial stretch between the two signal energies, whereas $\varphi(\theta;A)$ sets their relative angular displacement. The radial projections of the two energy spectra, defined as

$$P_n(\theta) \doteq \mathcal{RP}\big[\bar{\mathcal{E}}_n(\rho,\theta)\,|\,\rho\big] = \int_0^\infty \rho\bar{\mathcal{E}}_n(\rho,\theta)d\rho, \quad n = 1,2 \tag{4.80}$$

relate as

$$P_2(\theta) = \frac{\det(A)}{\lambda^2(\theta;A)}P_1\big(\varphi(\theta;A)\big), \quad 0 \le \theta < 2\pi. \tag{4.81}$$

Figures 4.4(a) and (b) show with solid blue lines the polar plots of the energy radial projections $P_1(\theta)$ and $P_2(\theta)$ relative to the images of Figures 4.2(a) and (e), respectively.

The estimate of the affine matrix A from Equation (4.81) can be set forth as a nonlinear minimization problem by representing the two radial projections $P_1(\theta)$ and $P_2(\theta)$ in terms of their Fourier series expansions [27]. Since they are periodic functions with period 2π, $P_1(\theta)$ can be approximated as

$$P_1(\theta) \doteq \tilde{P}_1(\theta) = \gamma_0 + \sum_{i=1}^{N} \gamma_i \cos(i\theta - \omega_i), \qquad (4.82)$$

where

$$\omega_i \doteq \arctan \frac{\beta_i}{\alpha_i}, \quad \gamma_i \doteq \frac{\alpha_i}{\cos \omega_i}, \quad \gamma_0 \doteq \frac{\alpha_0}{2}, \qquad (4.83a)$$

$$\alpha_i \doteq \frac{1}{\pi} \int_0^{2\pi} P_1(\theta) \cos(i\theta) \, d\theta, \quad i = 0, \cdots, N, \qquad (4.83b)$$

$$\beta_i \doteq \frac{1}{\pi} \int_0^{2\pi} P_1(\theta) \sin(i\theta) \, d\theta, \quad i = 1, \cdots, N. \qquad (4.83c)$$

The function $\tilde{P}_1(\theta)$ approximating the projection $P_1(\theta)$ in Figure 4.4(a) is shown with a dashed red line, which perfectly overlaps $P_1(\theta)$. The number of terms used in the Fourier series expansion should be tailored to the amount of small scale details (high frequencies) present in $P_1(\theta)$: the higher the frequencies in $P_1(\theta)$, the larger N should be. In the example of this section, we used $N = 100$, even though a smaller number of terms could have been used without any significant difference. A more efficient implementation would make N dependent on the degree of approximation required in Equation (4.82).

From Equation (4.81), matrix A can be estimated as $\hat{A}$ by solving the polar curve-fitting problem

$$\hat{A} = \arg\min_A \left\| \tilde{P}_2(\theta) - P_2(\theta) \right\|_{L_2([0,2\pi))}^2 = \arg\min_A \left\| \frac{\det(A)}{\lambda^2(\theta;A)} \tilde{P}_1(\varphi(\theta;A)) - P_2(\theta) \right\|_{L_2([0,2\pi))}^2 =$$

$$= \arg\min_A \left\| \frac{\det(A)}{\lambda_2(\theta;A)} \left(\gamma_0 + \sum_{i=1}^{N} \gamma_i \cos(i\varphi(\theta;A) - \omega_i) \right) - P_2(\theta) \right\|_{L_2([0,2\pi))}^2,$$

$$(4.84)$$

a.

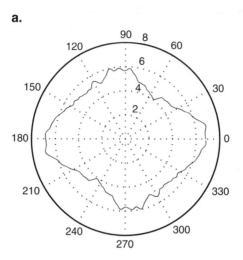

b.

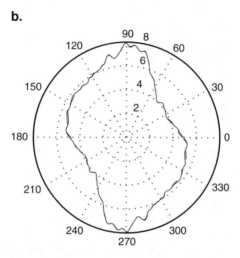

Figure 4.4 (a) Polar plots of the radial projection $P_1(\theta)$ [solid line] and of its approximation $\tilde{P}_1(\theta)$ [dashed line] relative to the image of Figure 4.2(a); (b) polar plots of the radial projection $P_2(\theta)$ [solid line] relative to the image of Figure 4.2(e) and of the affine-warped version $\tilde{P}_2(\theta)$ of $\tilde{P}_1(\theta)$ [dashed line].

i.e., by minimizing the distance, expressed as the squared $\mathcal{L}_2$-norm over $[0,2\pi)$, between $P_2(\theta)$ and the affine-warped version $\tilde{P}_2(\theta)$ of the approximation $\tilde{P}_1(\theta)$ of $P_1(\theta)$. This nonlinear minimization problem can easily be solved with the Levenberg-Marquardt algorithm [28] by using the rotation matrix, which makes the orientation [29] of $P_1(\theta)$ coincide with that of $P_2(\theta)$ as the initial estimate for A. Usually, fewer than twenty iterations are sufficient for the convergence of the Levenberg-Marquardt algorithm to the optimal solution.

The solution of the problem in Equation (4.84) delivers an estimate of the matrix A up to a sign reversal of its entries [26]. After compensating for the warping due to $\hat{A}$ in $\ell_2(x)$, the translational displacement b in Equation (4.73) can be estimated as $\hat{b}$ through a phase correlation algorithm, which also allows the discrimination between the two feasible solutions for A [20]. The values estimated by this algorithm for A and b are, respectively,

$$\hat{A} = \begin{bmatrix} 1.1966 & 0.2972 \\ -0.2017 & 0.9035 \end{bmatrix}$$

and $\hat{b} = 0$.

The polar curve $\tilde{P}_2(\theta) \doteq \left(\det(\hat{A}) / \lambda^2(\theta; \hat{A}) \right) \tilde{P}_1 \left(\varphi(\theta; \hat{A}) \right)$ associated with the estimate $\hat{A}$ is drawn with a dashed red line in Figure 4.4(b). It should be observed that $\tilde{P}_2(\theta)$ follows approximately the profile of $P_2(\theta)$; this is explained by the fact that the two radial projections are computed numerically over a finite circular region of the frequency domain centered at $k = 0$, whence minor differences in energy content may exist.

4.6   Conclusion

The frequency domain provides an insightful perspective on motion with useful applications to displacement estimation. This chapter provides an account of this fact relative to planar motion modeled by translations, rotations, and affine transformations. The robustness of the frequency domain methods for translational and rotation displacement estimation is worth pointing out. Current research addresses the frequency domain analysis of motion of 3-D rigid surfaces. A frequency domain technique for the estimation of 3-D rotations and translations, inspired by the above-mentioned research, has recently appeared in [30] and has been applied to the automatic construction of 3-D models [31]. Frequency domain methods for displacement estimation are currently being used in omnidirectional vision systems for robotics applications [32, 33, 34].

References

[1] Casasent, D. and Psaltis, D., "Position, Rotation, and Scale Invariant Optical Correlation," *Applied Optics*, Vol. 15, No. 7, pp. 1795–1799, 1976.

[2] Casasent, D., "Pattern Recognition: A Review," *IEEE Spectrum*, pp. 28–33, 1981.

[3] Peli, T., "An Algorithm for Recognition and Localization of Rotated and Scaled Objects," *Proceedings of the IEEE*, Vol. 69, No. 4, pp. 483–485, 1981.

[4] Zwicke, P.E. and Kiss, I., "A New Implementation of the Mellin Transform and Its Application to Radar Classification of Ships, *IEEE Transactions on Pattern Analysis and Machine Intelligence*, Vol. PAMI-5, pp. 191–198, 1983.

[5]   Sheng, Y. and Duvernoy, J., "Circular-Fourier-Radial-Mellin Descriptors for Pattern Recognition," *Journal of the Optical Society of America — A*, Vol. 3, No. 6, pp. 885–887, 1986.

[6]   Sheng, Y. and Arsenault, H.H., "Experiments on Pattern Recognition Using Invariant Fourier-Mellin Descriptors," *Journal of the Optical Society of America — A*, Vol. 3, No. 6, pp. 771–776, 1986.

[7]   Kretz, F. and Sabatier, J., "Échantillonage des Images de Télévision: Analyse dans le Domain Spatio-Temporel et dans le Domain de Fourier," *Ann. Télécommunications*, Vol. 36, pp. 231–273, 1981.

[8]   Watson, A.B. and Ahumada Jr., A.J., "A Look at Motion in the Frequency Domain," *NASA Technical Memorandum n. 84352*, NASA Ames Research Center, Moffett Field, CA, 1983.

[9]   Fleet, D.J. and Jepson, A.D., "Computation of Component Image Velocity from Local Phase Information," *International Journal of Computer Vision*, Vol. 5, No. 1, pp. 77–104, 1990.

[10] Cortelazzo, G.M. and Balanza, M., "Frequency Domain Analysis of Translations with Piecewise Cubic Trajectories," *IEEE Transactions on Pattern Analysis and Machine Intelligence*, Vol. 15, No. 4, pp. 411–416, 1993.

[11] Cortelazzo, G.M. and Nalesso, G., "A Differential Equation Approach to the Computation of the Fourier Transform of the Images of Translating Objects," *IEEE Transactions on Information Theory*, Vol. 40, No. 6, pp. 2049–2058, 1994.

[12] Cortelazzo, G.M., Balanza, M., and Monti, C., "Frequency Domain Analysis of Rotational Motion," *Multidimensional Systems and Signal Processing*, N.K. Bose, Editor, Vol. 4, pp. 203–225, Kluwer Academic Publishers, Boston, MA, 1993.

[13] Cortelazzo, G.M., Lucchese, L., and Monti, C., "Frequency Domain Analysis of General Planar Rigid Motion with Finite Duration," *Journal of the Optical Society of America — A*, Vol. 16, No. 6, pp. 1238–1253,1999.

[14] Lucchese, L. and Cortelazzo, G.M., "Frequency Domain Analysis of Affine Motion with Linear Time Dependence and Finite Duration," *Proceedings of the International Conference on Signal Processing and Communications* (ICSPC'98), Gran Canaria, Canary Islands, Spain, pp. 28–31, 1998.

[15] Kuglin, C.D. and Hines, D.C., "The Phase Correlation Image Alignment Method," *Proceedings of the IEEE 1975 International Conference on Cybernetics and Society*, pp. 163–165, 1975.

[16] De Castro, E. and Morandi, C., "Registration of Translated and Rotated Images Using Finite Fourier Transforms," *IEEE Transactions on Pattern Analysis and Machine Intelligence*, Vol. PAMI-9, No. 5, pp. 700–703, 1987.

[17] Alliney, S., "Digital Analysis of Rotated Images," *IEEE Transactions on Pattern Analysis and Machine Intelligence*, Vol. PAMI-15, No. 5, pp. 499–504, 1993.

[18] Reddy, B.S. and Chatterji, B.N., "An FFT-Based Technique for Translation, Rotation, and Scale-Invariant Image Registration," *IEEE Transactions on Image Processing*, Vol. IP-5, No. 8, pp. 1266–1271, 1996.

[19] Alliney, S., Cortelazzo, G.M., and Mian, G.A., "On the Registration of an Object Translating on a Static Background," *Pattern Recognition*, Vol. 29, No. 1, pp. 131–141, 1996.

[20] Lucchese, L. and Cortelazzo, G.M., "A Noise-Robust Frequency Domain Technique for Estimating Planar Roto-Translations," *IEEE Transactions on Signal Processing*, Vol. 48, No. 6, pp. 1769–1786, 2000.

[21] Bracewell, R.N., Chang, K.-Y., Jha, A.K., and Wang, Y.-K., "Affine Theorem for Two-Dimensional Fourier Transform," *Electronics Letters*, Vol. 29, No. 3, p. 304, 1993.

[22] Bracewell, R.N., *The Fourier Transform and Its Applications*, McGraw-Hill, New York, 1965.

[23] Krumm, J. and Shafer, S.A., "Shape from Periodic Texture Using the Spectrogram," *Proceedings of the International Conference on Computer Vision and Pattern Recognition* (CVPR'92), Urbana-Champaign, IL, pp. 284–289, 1992.

[24] Malik, J. and Rosenholtz, R., "Computing Local Surface Orientation and Shape from Texture for Curved Surfaces," *International Journal of Computer Vision*, Vol. 23, No. 2, pp. 149–168, 1997.

[25] Krüger, S.A. and Calway, A.D., "A Multiresolution Frequency Domain Method for Estimating Affine Motion Parameters," *Proceedings of the International Conference on Image Processing* (ICIP'96), Lausanne, Switzerland, Vol. I, pp. 113–116, 1996.

[26] Lucchese, L., Cortelazzo, G.M., and Monti, C., "Estimation of Affine Transformations Between Image Pairs via Fourier Transform," *Proceedings of the International Conference on Image Processing* (ICIP'96), Lausanne, Switzerland, Vol. 3, pp 715–718, 1996.

[27] Lucchese, L., "A New Method for Perspective View Registration," in *Proceedings of the International Conference on Image Processing* (ICIP'00), Vancouver, BC, Canada, Vol. 2, pp. 776–779, 2000.

[28] Press, W.H., Flannery, B.P., Teukolsky, S.A., and Vetterling, W.T., *Numerical Recipes in C: The Art of Scientific Computing*, 2nd Ed., Cambridge University Press, Cambridge, UK, 1992.

[29] Horn, B.K.P., *Robot Vision*, MIT Press, Cambridge, MA, 1986.

[30] Lucchese, L., Doretto, G., and Cortelazzo, G.M., "A Frequency Domain Technique for 3D View Registration," *IEEE Transactions on Pattern Analysis and Machine Intelligence*, Vol. 24, No. 11, pp. 1468–1484, 2002.

[31] Andreetto, M., Brusco, N., and Cortelazzo, G.M., "Automatic 3D Modeling of Textured Heritage Objects," *IEEE Transactions on Image Processing*, Vol. 13, No. 3, pp. 354–369, 2004.

[32] Makadia, A. and Daniilidis, K., "Direct 3D-Rotation Estimation From Spherical Images Via a Generalized Shift Theorem," *Proceedings of the International Conference on Computer Vision and Pattern Recognition* (CVPR'03), Madison, WI, Vol. 2, pp. 217–224, 2003.

[33] Ishiguro, H. and Tsuji, S., "Image-Based Memory of Environment," *Proceedings of the IEEE/RSJ International Conference on Intelligent Robots and Systems* (IROS-96), pp. 634–639, 1996.

[34] Menegatti, E., Zoccarato, M., Pagello, E., and Ishiguro, H., "Hierarchical Image-Based Localisation for Mobile Robots With Monte-Carlo Localisation," *Proceedings of the European Conference On Mobile Robots* (ECMR'03), Warsaw, Poland, pp. 13–20, 2003.

chapter 5

Quality of service assessment in new generation wireless video communications

Gaetano Giunta

Contents

5.1   Introduction

Third generation (3G) communications expand the capability to carry several customers at the same time, while offering innovative services with higher bit rates for multimedia applications [1]. In addition to being employed for vocal links, mobile phones will develop into multifunctional multimedia terminals. 3G cellular technologies will bear bit rates up to 384 kbps outdoors and 2 Mbps indoors [2]. Furthermore, new generation, faster indoor wireless standards, such as Bluetooth (802.15), WLAN (802.11), and ultra wideband, will be used [3]. Further types of applications will be employed for mobile devices by means of programmable digital signal processors [4]. It will be feasible to manage compressed video, including graphics and images, together with audio or speech with low-power processors, available to execute hundreds of operations per second [1].

Efficient compression techniques are necessary for even the higher 3G bit rates [5]. As an example, a standard TV 24-bit color video with 640×480 pixels at 30 frames per second requires 221 Mb/s. Compression tools make it possible to support multimedia contents, often at a reduced resolution. Perspective standardization of compression algorithms should allow interoperability among terminals from different companies. However, many standards exist for several applications, depending on system resources, content type, and the quality required by the user. Moreover, popular non-standard formats are widely employed, involving flexible devices and adaptive programming.

Internet compatibility, market differentiation, and new lifestyles are the major reasons for adding multimedia capabilities. A communication device with an embedded signal processor may constitute a supplementary apparatus to add multimedia capability and, as a consequence, value to products. Mobile phones are now able to access simplified Internet pages by the WAP protocol. Wireless communications devices can certainly enrich the mobile lifestyle through added video and audio capabilities [1].

Possible mobile video applications include streaming video players, videophones, video e-postcards and messaging, surveillance, and telemedicine [3]. In the case of short non-real-time multimedia services, such as for video e-postcards or messages, data can be buffered. For surveillance and telemedicine, a sequence of high-quality still images, or low-frame-rate video, may be necessary. With streaming video, off-line encoding is possible. The necessity of streaming decoding while data are received is somewhat comparable to conventional TV, but with mobile video the decoding is much cheaper and the results are of poorer quality [6]. For one-way decoding, some buffering and delay are tolerable. Conversely, videophone applications require simultaneous encoding and decoding, with limited delay, resulting in further quality reduction. In fact, the two-way videophone is perhaps the most difficult to implement among all the mobile video applications [1].

Video coding presents relevant differences in comparison with speech coders, due to the increased bandwidth and the dynamic range, as well as higher dimensionality of data. This has led to the use of variable bit rate, predictive, error-sensitive, lossy compression, and standards that are not bit-exact [1]. For video transmitted over a wireless channel, it may happen that the decoder does not successfully remove particular error patterns; this may disseminate the error along to the following frames [7]. While the variability in terms of bit rate is crucial to achieving higher compression ratios, a bounded bit rate must be maintained to avoid buffer overflows, matching to the available channel capacity. In such a case, either pixel precision or frame rate must be adjusted dynamically, determining that the effective quality may vary within or between the frames [6].

5.2 Wireless multimedia communication trends and services

Universal mobile telecommunications system (UMTS) services will not only offer mobile services supported by second generation systems such as GSM but will also expand these services to higher rates and greater flexibility [8]. UMTS services are supporting user transmission rates that can dynamically vary up to 2 Mbps. The highest rates will be mainly employed in indoor environments. Nevertheless, there will be substantial increases in rates throughout all environments, in comparison with the typical 9.4 kbps of GSM. The future trend is going to provide the transmission rates required by most potential users, but the practical question arises of what type of services can be performed, as well as when and where. There exists a clear trend for the convergence of IP protocol to wireless, that is, what we now call *wireless IP*, leading to the *wireless Internet* [8]. Innovative multimedia services are going to generate the need for new service enablers as well as new data processing and managing techniques.

The four main classes of UMTS traffic consist of conversational, streaming, interactive, and background. They are differentiated by their delay sensitivity. Conversational classes have higher delay sensitivity than background classes. The first two classes correspond to real-time classes, unlike the latter ones. Typical speech over circuit-switched bearers, voiceover IP (VoIP), and video telephony represent the (real-time) conversational class. Video telephony is based on symmetric traffic with end-to-end delay thresholds below 399 ms. Video telephony has higher bit-error-rate requirements than speech due to its video compression features; however, it has the same delay sensitivity of speech [8]. UMTS recommends ITU-T Rec. H.324M (the suffix M stands for mobile) for video telephony in circuit-switched links [9]. Two video telephony options exist for packet-switched links, i.e., ITU-T Rec. H.323 [10] and TF SIP [11]. The mobile-adapted H.324 includes fundamental elements such as H.223 for multiplexing and H.245 for control, as well as H.263 video codec, G.723.1 speech codec, and V.8bis. MPEG-4 video and AMR speech coding (i.e. the CELP-based standard, the same as in GSM) can be used optionally to better suit UMTS services. Technical specifications include seven phases for a call, i.e., setup, speech only, modem learning, initialization, message, end, and clearing. Backward compatibility occurs through level 0 of the H.223 multiplexing, which is the same as H.324. The H.324 terminal also has an operation mode for use over ISDN links [12]. The UMTS call control mechanism takes into account V.8bis messages. Basic V.8bis features incorporate the following capabilities [8]: flexible communication mode selection by either the calling or answering party; enabling automatic identification of common operating modes; enabling automatic selection among multiple terminals sharing common telephone channels; and friendly user interface to switch from voice telephony to modem-based communications.

The H.323 has similar characteristics to H.324M. The H.323 ITU-T protocol standard for multimedia call control allows packet-switched multimedia communications in UMTS. It involves gateways and terminals to provide call control and processing functions. Moreover, the standard provides multiple options for voice, data, and video communications and may employ a gatekeeper function to provide resource management functions [8]. As an international standard [10] for teleconferencing over packet networks, the H.323 acts as a single standard to permit Internet telephony products to interoperate; it also has the flexibility to support different HW/SW and network capabilities. When both audio and video media act in a conference, they transmit using separate Real-time Transport Protocol (RTP) sessions. In fact, the packets are transmitted using two different User Datagram Protocol (UDP) port pairs and/or multicast addresses. Thus, direct coupling does not exist at the RTP level between audio and video sessions, and synchronized playback of a source's audio and video takes place using timing information carried in the data packets for both sessions [8].

The Session Initiation Protocol (SIP) [11] is another alternative to enable packed-switched video telephony. Such protocol is an application layer control signaling protocol for creating, modifying, and terminating sessions with one or more participants, e.g., Internet multimedia conferences, Internet telephone calls, and multimedia distribution. Communication is either multicast or unicast. Technically, SIP protocol has the following characteristics [8]: called and calling users can choose the preferred address for the connection; users can use a typical e-mailing address with username and domain; it is orthogonal to other signaling protocols; it uses servers for redirection, user location tracking, and proxy capabilities; it does not have address initiation and termination like H.323; and it is simple and easy to implement by Internet developers.

The streaming technique by SIP protocol assists Internet browsing by allowing partial displays of the transferred information [8]. It supports the large asymmetry of Internet applications. Through buffering, the streaming technique smoothes out packet traffic and supports video on demand and web broadcast. Both types of video applications use the same compression technology but they differ in the provided services, depending on the transmission rate or delay sensitivity of the network. In fact, in the context of Internet applications like web browsing, the response time will depend on the type of information requested and the quality of the link as well as protocols in use. Conversely, delay-sensitive applications will demand faster interaction. Other applications, such as location services, games, and passive information centers, will operate within flexible round-trip delays.

Meeting needs implies making available the correct tools and environment [8]. In fact, a new generation mobile phone not only needs to be a smart device capable of accessing a packet-switched network, support bandwidth on demand, audio streaming, and multimedia, it will also need versatility and multiple capabilities. A multifunctional device will make the difference in future usage and acceptance of higher transmission rates offered through UMTS [3, 8]. Market penetration and widespread usage of these multimedia services will depend on the available and affordable terminals as well as the pragmatic applications [9]. Wireless device interconnections, intelligent voice recognition, wireless e-mail, simultaneous voice and data, user-defined closed user groups, location services [13, 14], personal profile portals, and location-based delivery and marketing will occur only with efficient integration and interworking of multiple technologies. Applications may not necessarily come from the technology design; the final blend will depend on the available and accessible technology [8]. Therefore, the capacity of creating innovative services is correlated to the quality of the same service that the provider (and the link) is able to guarantee.

5.3 Quality of service requirements

This problem of "matching" the application to the network is often described as a quality of service (QoS) problem [15]. The problem is twofold: there is the QoS required by the application, which relates to visual quality perceived by the user, and the QoS reachable by the transmission channel and the network, which depends on the physical link's capabilities. Most of the encoding algorithms for real-time video are lossy; i.e., some distortion is introduced by encoding [16, 17]. In practice, the amount of distortion should be adequate for the specific application. Its acceptability depends on the communication and displaying scenario. Distortion should be near constant from a subjective viewpoint. Sudden changes in quality (such as "blocking" effects due to rapid motion or distortion due to transmission errors) can have a very negative effect on perceived quality from the viewer. In fact, the viewer will quickly become accustomed to a given video quality level.

The key parameters for QoS requirements about transmission are the mean bit rate and the variation of the bit rate [15]. The mean rate depends on the compression algorithm as well as the characteristics of the source video (such as frame size, number of bits per sample, frame rate, and amount of motion). Many video-coding techniques include a certain degree of compression tuning that bound the mean rate after encoding. However, the achievable mean compressed bit rate is actually limited for a given source (with a particular frame size and frame rate). For example, "broadcast TV quality" video (approximately 704 × 576 pixels per frame, 25 or 30 frames per second) encoded using MPEG-2 requires a mean encoded bit rate of around 2–5 Mbps for acceptable visual quality [18, 19]. The network or channel must support (at least) such a bit rate in order to successfully transmit video at broadcast quality.

Real-time video is sensitive to delay by nature. The QoS requirements in terms of delay depend on whether video is transmitted one way (e.g., broadcast video, streaming video, playback from a storage device) or two ways (e.g., video conferencing) [15]. The original temporal relationships between frames have to be preserved in simplex (one-way) video transmission because any associated media that is "linked" to the video frames should remain synchronized, but a certain degree of congruent delay is admitted. The most common example is accompanying audio, where a loss of synchronization of more than about 0.1 s can be obvious to the viewer [15]. Duplex (two-way) video transmission has the above constraints (constant delay in each direction, synchronization among related media) plus the obligation that the total delay from acquisition to display must be low enough. A rule of thumb for video conferencing is to keep the total delay less than 0.4 s [15]. If the delay is longer than this, video conversation will appear nonnatural. Interactive applications also have a requirement of low delay. In such a case, a delay that is too long between the user's action and the corresponding effect on the video source makes the application unresponsive.

5.4   QoS assessment in a cellular system

Cellular networks grow in size and complexity. For this reason, automated network management is very appealing. As a consequence, there is a need for simple or automated operation and maintenance of third-generation networks, to decrease operators' costs and to use hardware investments in the most effective way. A real-time, performance-monitoring application provides immediate feedback on the performance of the network, providing a flexible system based on real-time events instead of on counter-based statistics [20]. Operators can thus respond promptly to eventual problems detected in the network and see the immediate effect of configuration updates. This gives almost unlimited possibilities of finding specific behaviors in the network. It is useful for troubleshooting as well as for introducing and tuning new features or predicting possible changes in the parameter settings. Real-time performance events could also act as predefined reports of a wide range of performance-monitoring services that depend on the correlation of events; such services may include network traffic supervising, quality indices, and, finally, the identification of specified situations in the network. Further examples include identifying cells with a large amount of fast-moving mobile stations (with possibly ping-pong handover situations) and the monitoring of end-user performance to determine whether the customer gets what he is paying for.

Using powerful compression algorithms while maintaining good visual quality has been the priority in the design of new digital video systems, and guaranteeing a certain level of quality is an important concern for content providers. Variations in quality are due to lossy compression as well as to transmission errors, both of which lead to artifacts in the received material. The accurate measurement of quality as perceived by the user has become one of the great challenges because the amount and visibility of these distortions strongly depend on the actual video content. Several benchmark procedures for subjective experiments have been formalized in ITU-R Recommendation BT.500 [21], which suggests the following: standard viewing conditions; criteria for the selection of observers; and test material, assessment procedures, and data analysis methods. The problem with subjective experiments is that they are time consuming, expensive, and often impractical [22].

For a lot of applications, such as online quality monitoring and control, engineers have conversely turned to simpler error measures, such as mean squared error (MSE) or peak signal-to-noise ratio (PSNR), suggesting that they would be equally valid [22]. However, these simple measures operate solely on the basis of pixel-wise differences and, therefore, their predictions often do not agree well with perceived quality. Another way to perform objective measurements of data transmission is looking at network-related parameters such as bit error rate (BER) or packet loss ratio (PLR). Establishing and maintaining a certain level of network QoS for several applications is a very active research area at the moment [23, 24]. Unfortunately, once again the measurements and the employed error-control protocols have no direct relation to the video quality as perceived by the end user.

The weakness of these methods has led to the study of different perceptual quality measures. In principle, two different approaches can be distinguished, namely methods based on models of the human visual system models and techniques exploiting *a priori* knowledge about the compression and transmission [25, 22]. Examples of the former metrics (potentially most accurate but computationally expensive due to their complexity) are described in [26, 27]. Examples of the latter specialized metrics using ad hoc techniques include [28, 29]. While such metrics are not as versatile, they normally perform well in a given application area [22]. In practice, their main advantage lies in the fact that they often allow a computationally more efficient implementation. Several of these video quality metrics were compared against subjective ratings for a well-defined set of test sequences by the Video Quality Experts Group (VQEG), as described in [30, 31] in more detail.

Unlike out-of-service testing, in which the full reference video is available to the metrics, in-service metrics are designed to monitor and control systems while they are in operation [22]. They can be used to carry out measurements at practically any point of the transmission chain. This is a particularly important issue in multimedia-streaming applications. The setup can be intrusive or not, depending on the objective of the test and the nature of the testing methodology. The corresponding metrics are often referred to as reduced-reference and no-reference metrics, respectively [22]. The fact that the full reference video is usually not available for comparison makes an accurate assessment much more difficult for in-service metrics. The algorithms are generally based on some *a priori* knowledge about the scene content, the compression method, and/or the expected artifacts [22]. Most of them (e.g., see [32, 33]) aim at identifying certain features in a scene and assessing their distortion. They also take into account knowledge of the compression method and the corresponding artifacts (for details on the typical compression artifacts, see [34]).

5.5   *Tracing watermarking for QoS assessment*

5.5.1   *Motivation and operating target*

The original proposition of using watermarking for automatic quality monitoring is quite recent [35]. In practice, the watermark (narrow-band low-energy signal) is spread over the image (larger bandwidth signal) so that the watermark energy contribution for each host frequency bin is negligible, which makes the watermark nearly imperceptible. This basic idea was applied by the author of this chapter (together with the components of the Signal Processing and Multimedia Communication Group at the University of Roma Tre) to evaluate the quality of service in multimedia communications by means of an unconventional use of fragile watermarking [36, 37, 38]. These papers (as well as further unpublished material) are here abstracted as the key contributions on the method, and the results are reported in this section.

The rationale behind the approach is that the alterations suffered by the watermark are likely to be suffered by the data, since they follow the same communication link; therefore, the watermark degradation can be used to evaluate the alterations endured by the data, as shown in the principle scheme of Figure 5.1. At the receiving side, the watermark is extracted and compared with its original counterpart. Spatial spread-spectrum techniques [39, 40] perform the watermarking embedding.

The core of this work is the proposal of an optimum operating procedure, involving the definition of a number of QoS levels and a blind assignment of the received multimedia stream to one of the possible classes of service. In practice, the receiving mobile station is able to verify whether the negotiated quality level has been actually maintained by the effective link or not. In such a case, the operator can lower the user's fare and bustle about in order to guarantee an improved service level by modifying the operating settings of the communication link. In the simplest version, the two hypotheses consist of "good" and "not good" quality levels.

The blind operating system at the receiver of a communication link needs to perform such a hypothesis test in order to match the negotiated quality level and the effectively provided one. If the user had requested a secure (high-level) transmission, he has to be charged accordingly. Otherwise, the connection can be low-priced, even free of charge. The economic impact of a decision error is dramatic for the operator. In fact, a false alarm error can have a tremendous cost for the operator because it may allow a user to effectively transmit and receive for free. The statistical occurrence rate of such a false alarm error should be limited. In practice, the maximum allowed rate could be determined by a proper analysis of the economic costs and benefits.

Conversely, an erroneous decision can directly affect the customer who pays too much for a low-quality call when the user had requested high-level transmission. This decision has no direct costs for the operator. In fact, it

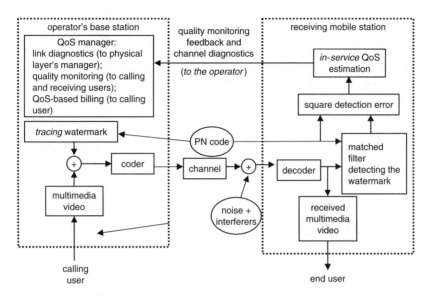

Figure 5.1 Principle scheme of the quality monitoring system for wireless multimedia communications based on tracing watermarking.

continues to charge the multimedia call at the negotiated higher price. In such a case, the typical user drops the call line and dials again. Service inefficiency provided to the end user risks pushing the customer to discontinue using multimedia services from that operator. Moreover, the operator must be very careful when determining the probability of such an event, since service efficiency is one of the parameters considered by customers when they choose one operator over others. As a consequence, the probability of such an event must be minimized by the operating system.

The economic costs should also include the computational complexity of signal processing running on mobile terminals. Mobile stations are expected to host, in the near future, large processing capabilities because of the monotonically decreasing cost of very large scale integration. In fact, several tasks could be accomplished by mobile terminals, such as onboard adaptive beamforming (spatio-temporal array processing) and near-optimum multiuser rake reception (suggested to be employed even in downlink) (e.g., see [41]). Moreover, in a wireless video communication scenario like UMTS, the coded video sequence has to be decoded at the receiving side. Although actual mobile phones do not perform hardware and software processing, the authors believe that, in a few years, the complexity of the QoS evaluation procedure presented in this paper will appear negligible in comparison with MPEG-2/4 decoding and adaptive onboard array processing.

As discussed in [36], it can be observed that in real-time systems requiring error control, such as video telephony and streaming, a number of consequences follow a poor quality decision from the mobile station: first, the bit rate (and the video resolution) transmitted by the base station is lowered in a few seconds (by a higher spreading factor in the UMTS systems); second, if the mobile station declares a null quality, the link is broken down by the operator; and third, frequent declarations of poor or null quality will be a reason for the operator's admission call manager to refuse the user access to making further calls for a given time period (until the mobile station moves into a region with higher SNR). For these reasons, the mobile station should avoid fraudulent declarations of bad quality in order to transmit free of charge. In fact, the result of a possible false declaration about QoS is a sudden cut in communication and the risk of not allowing further calls.

As shown in [36], it is effectively possible to blind-estimate the quality of a transmission system (including the coder quality) without affecting the quality of the video communications. In particular, such a QoS index can be usefully employed for a number of different purposes in wireless multimedia communication networks (see again Figure 5.1), including control feedback to the sending user on the effective quality of the link, detailed information to the operator for billing purposes, and diagnostic information to the operator about the communication link status.

5.5.2   Signal processing algorithm and results

A set of uncorrelated pseudo-random noise (PN) matrices (one per frame and known to the receiver) is multiplied by the reference watermark (one for the entire transmission session and known to the receiver). Like in spread-spectrum techniques, the use of different spreading PN matrices assures that the spatial localization of the mark is different frame by frame so that the watermark visual persistency is negligible. Moreover, the method is robust against permanent bit errors, due to either the physical network or its management (e.g., multipath of the transmission channel, multiuser interference, and excess loading factors). After generating the marks, the embedding of the tracing mark is performed in the Discrete Cosine Transform (DCT) Domain [39, 42, 43]. The two-dimensional watermark is embedded in the DCT middle-band frequencies of the whole luminance frame of the transmitted video, as depicted in Figure 5.2. The watermark is randomized by the PN matrices and added to the DCT of each frame. After the inverse DCT, the whole sequence is coded by a video coder (either MPEG2 or MPEG4 coders have been used in [36]) and transmitted.

In practice, the watermarks are embedded in a (large) number M of the luminance I-frames. In fact, any nth ($n=1..M$, with $M>>1$) I-frame does not suffer from any interpolation error, unlike the P-frames and B-frames. In this way, the watermark is affected only by the channel's errors, and at the receiver side, the estimation of the degradations affecting the received mark can be used to provide quality assessment of the channel. The result is that the spatial localization of the mark is different frame by frame so that the visual latency perception is negligible yet robust against permanent bit errors, due both to the physical network and its management (e.g., multipath of the transmission channel, multiuser interference, and excess loading fac-

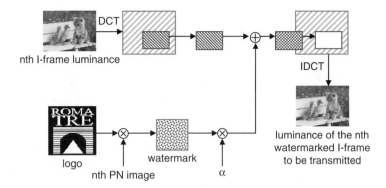

Figure 5.2 Block scheme of the watermark insertion in a transmitted video.

tors). In other words, the watermark is randomized by the PN matrices and added to the DCT of each frame. The mark is multiplied by a scale factor <<1 to properly reduce the watermark power; the visibility of the mark in the video sequence increases proportionally to this factor.

The receiver implements video decoding as well as watermark detection. In fact, at the same time after decoding of the video stream, a matched filter extracts the (known) watermark from the DCT of each *n*th received luminance I-frame of the sequence (see Figure 5.3). The estimated watermark is matched to the reference one (despread with the *known* PN matrix). The matched filter is tuned to the particular embedding procedure so that it can be matched to the randomly spread watermark only. It is assumed that the receiver knows the initial spatial application point of the mark in the DCT domain. The QoS is evaluated by comparing the extracted watermark with respect to the original one (see again Figure 5.1). In particular, its mean square error (MSE) is evaluated as an index of the effective degradation of the provided QoS. A possible index of the degradation is simply obtained by calculating the mean of the error energy:

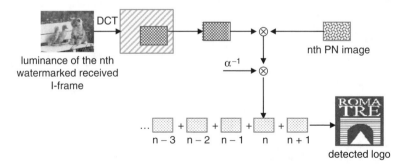

Figure 5.3 Block scheme of the watermark detection in a received video.

$$E_e = \frac{1}{M} \sum_{n=1}^{M} \left(\sum_{k_1=1}^{K_1} \sum_{k_2=1}^{K_2} (w_n[k_1,k_2] - \hat{w}_n[k_1,k_2])^2 \right)$$

where $w_n[k_1,k_2]$ and $\hat{w}_n[k_1,k_2]$, represent the original watermark and the extract watermark, respectively, and $n=1..M$ is the current frame index.

Several simulations have been made to detect the sensitivity of the digital watermark to two different degradation sources (the noisy channel affects the video quality; the co-decoder itself affects the perceived image quality). Detailed results, obtained using either MPEG-2 or MPEG-4, can be found in [36, 37, 38]. In practice, the decoding quality depends on three factors: 1) the multiplying factor of the quantization matrix: as it increases, the video quality also decreases; 2) the bit rate: as it increases, it improves the video quality; and 3) the number of intracoded frames. To evaluate the sensibility of the method to the degradations due only to the coder, tests have been performed with ideal channel and different coder quality. In particular, Figure 5.4 shows the MSE of the decoded QCIF (174×144 pixels) "Miss America" video (upper curve) and the watermark detected from 50 I-frames (lower curve) versus the (variable) bit rate of the employed MPEG-2 coder. Moreover, the MSEs are depicted in Figure 5.5 for the same examples of application, versus the DCT quantization index used in the MPEG-2 algorithm to tune the final image quality. The above trials confirm that the watermark error is roughly proportional to the coding quality, which conversely depends on the available channel capacity.

From the operating viewpoint, we are interested in the actual video over a given communication link determined by the (booked) maximum bit rate. In other words, the target quality is fixed by the negotiated channel capacity

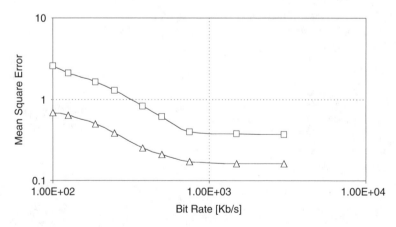

Figure 5.4 Mean square error of the decoded "Miss America" video (squares) and the detected watermark (triangles) versus the (variable) bit rate of an MPEG-2 coder.

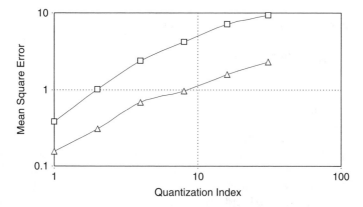

Figure 5.5 Mean square error of the decoded "Miss America" video (squares) and the detected watermark (triangles) versus the (variable) DCT coding quantization index.

while the actual quality depends also on the symbol errors introduced in the received data stream by the physical link, due to background noise as well as multipath channel and interference effects. Such transmission errors are modeled as a random Poisson process, characterized by a parametric probability of symbol error (proportional to the bit error rate). Then, the actual quality needs to be detected by an in-service quality assessment measure, such as the tracing watermarking technique presented in this subsection. For this purpose, a number of simulations have been carried out. A number of results can be found in [36, 37, and 38]. Figure 5.6 reports the

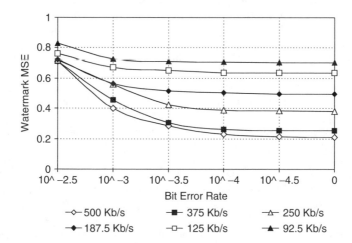

Figure 5.6 Mean square error of the watermark detected from the "Miss America" sequence versus the bit error rate of a simulated link for several (target) bit rates.

mean square error of the watermark detected from 50 I-frames of the MPEG-2 "Miss America" sequence versus the bit error rate of a simulated link for several (target) bit rates. The obtained results evidence the sensitivity of the watermarking quality index to the actual quality for given target quality levels.

5.6 Conclusion

In this chapter, motivation and operating guidelines for quality of service assessment in new generation wireless video communications have been discussed and investigated. In particular, the issue has been introduced in Section 5.1, and wireless multimedia trends and services are summarized in Section 5.2. In Sections 5.3 and 5.4, the QoS requirements of communication networks, as well as the guidelines for QoS assessment in cellular systems, are respectively reported. Motivations and operating targets of a recent method for QoS assessment based on tracing watermarking, together with the signal-processing algorithm and the numerical results, are finally reported in Section 5.5.

References

[1] Webb, J. and Lueck, C.: Video and Audio Coding for Mobile Applications. In: A. Gatherer and E. Auslander, eds, *The Application of Programmable DSPs in Mobile Communications*, John Wiley & Sons, New York, 2002.

[2] 3GPP website, http://www.3gpp.org.

[3] Bi, Q., Zysman, I., and Menkes, H., Wireless Mobile Communications at the Start of the 21st Century, *IEEE Communications Magazine*, pp. 110–116, 2001.

[4] Budagavi, M. and Talluri, R., Wireless Video Communications. In: J. Gibson, ed, *Mobile Communications Handbook*, 2nd ed., CRC Press, Boca Raton, FL, 1999.

[5] Budagavi, M., Heinzelman, W.R., Webb, J., and Talluri, R., Wireless MPEG-4 Video Communication on DSP Chips, *IEEE Signal Processing Magazine*, pp. 36–53, 2000.

[6] Dixit, S., Guo, Y., and Antoniou, Z., Resource Management and Quality of Service in Third Generation Wireless Networks, *IEEE Communications Magazine*, pp. 125–133, 2001.

[7] Talluri, R., Error-Resilient Video Coding in the ISO MPEG-4 Standard, *IEEE Communication Magazine*, vol. 36, pp. 112–119, 1998.

[8] Castro, J.P., *The UMTS Network and Radio Access Technology: Air Interface Techniques for Future Mobile Systems*, John Wiley & Sons, New York, 2001.

[9] Technical Specification Group, Codec for Circuit Switched Multimedia Telephony Service, General Description, 3GPP, TS 26 110, 1999.

[10] ITU-T H.323, Packet Based Multimedia Communications Systems, 1998.

[11] Handley, M. et al. SIP: Session Initiation Protocol, RFC2543, IETF, 1999.

[12] ITU-T H.324, Terminal for Low Bit-rate Multimedia Communication, 1998.

[13] 3GPP, Technical Specification Group (TSG) RAN, Working Group 2 (WG2), Stage 2, Functional Specification of Location Services in URAN, 3G TR 25.923, 1999.

[14] 3GPP, Technical Specification Group Services and System Aspects, Services and System Aspects, Location Services (LCS), Service description, Stage 1, 3G TS 22.071, 1999.

[15] Richardson, I.E.G., *Video Codec Design*, John Wiley & Sons, New York, 2002.

[16] Girod, B. and Farber, N., Error-Resilient Standard-Compliant Video Coding. In: A. Katsaggelos and N. Galatsanos, eds, *Recovery Techniques for Image and Video Compression*, Kluwer Academic, Dordrecht, 1998.

[17] Wang, Y., Wenger, S., Wen, J., and Katsaggelos, A., Review of Error Resilient Coding Techniques for Real-Time Video Communications, *IEEE Signal Processing Magazine*, vol. 17, pp. 61–82, 2000.

[18] Tsekeridou, S. and Pitas, I., MPEG-2 Error Concealment Based on Block-Matching Principles, *IEEE Transactions on Circuits and Systems for Video Technology*, vol. 10, pp. 646–658, 2000.

[19] Zhang, J., Arnold, J. F., and Frater, M., A Cell-Loss Concealment Technique for MPEG-2 Coded Video, *IEEE Transactions on Circuits and Systems for Video Technology*, vol. 10, pp. 659–665, 2000.

[20] Gustås, P., Magnusson, P., Oom, J., and Storm, N., Real-Time Performance Monitoring and Optimization of Cellular Systems, *Ericsson Review*, no. 1, pp. 4–13, 2002.

[21] ITU-R Recommendation BT.500-10: Methodology for the Subjective Assessment of the Quality of Television Picture, ITU, Geneva, Switzerland, 2000.

[22] Winkler, S., Sharma, A., McNally, D., Perceptual Video Quality and Blockiness Metrics for Multimedia Streaming Applications, *Proceedings of Wireless Personal Multimedia Communications*, WPMC01, vol. 2, pp. 553–556, 2001.

[23] Mishra, A., *Quality of Service in Communications Networks*, John Wiley & Sons, New York, 2001.

[24] Wang, Z., *Internet QoS: Architectures and Mechanisms for Quality of Service*, Morgan Kaufmann, San Francisco, 2001.

[25] Winkler, S., Issues in Vision Modeling for Perceptual Video Quality Assessment, *Signal Processing*, vol. 78, no. 2, pp. 231–252, 1999.

[26] Lubin, J. and Fibush, D., Sarnoff JND Vision Model, T1A1.5 Working Group, Document 97-612, ANSI T1 Standards Committee, 1997.

[27] Winkler, S., A Perceptual Distortion Metric for Digital Color Video, Proceedings of SPIE, vol. 3644, pp. 175–184, 1999.

[28] Tan, K. T., Ghanbari, M., and Pearson, D. E., An Objective Measurement Tool for MPEG Video Quality, *Signal Processing*, vol. 70, no. 3, pp. 279–294, 1998.

[29] Watson, A. B. et al., Design and Performance of a Digital Video Quality Metric, *Proceedings of SPIE*, vol. 3644, pp. 168–174, 1999.

[30] Rohaly, A. M. et al., Video Quality Experts Group: Current Results and Future Directions, *Proceedings of SPIE*, vol. 4067, pp. 742–753, 2000.

[31] VQEG, Final Report from the Video Quality Experts Group on the Validation of Objective Models of Video Quality Assessment, 2000, available at http://www.its.bldrdoc.gov/vqeg.

[32] Bretillon, P. and Baina, J., Method for Image Quality Monitoring on Digital Television Networks, *Proceedings of SPIE*, vol. 3845, pp. 298–307, 1999.

[33] Wolf, S. and Pinson M. H., Spatial-Temporal Distortion Metrics for In-Service Quality Monitoring of Any Digital Video System, *Proceedings of SPIE*, vol. 3845, pp. 266–277, 1999.

[34] Yuen, M. and Wu, H. R., A Survey of Hybrid MC/DPCM/DCT Video Coding Distortions, *Signal Processing*, vol. 70, no. 3, pp. 247–278, 1998.

[35] Holliman, M. and Yeung, M. M., Watermarking for Automatic Quality Monitoring, *Proceedings of SPIE, Security and Watermarking of Multimedia Contents,* vol. 4675, pp. 458–469, 2002.

[36] Campisi, P., Carli, M., Giunta, G., and Neri, A., Blind Quality Assessment System for Multimedia Communication Using Tracing Watermarking, *IEEE Transactions on Signal Processing,* vol. 51, no. 4, pp. 996–1002, 2003.

[37] Campisi, P., Carli, M., Giunta, G., and Neri, A., Tracing Watermarking for Multimedia Communication Quality Assessment, *IEEE International Conference on Communications,* New York, vol. 2, pp. 1154–1158, 2002.

[38] Campisi, P., Giunta, G., and Neri, A., Object Based Quality of Service Assessment for MPEG-4 Videos Using Tracing Watermarking, *IEEE International Conference on Image Processing,* Rochester, NY, vol. 2, pp. II.881–II.884, 2002.

[39] Cox, I., Kilian, J., Leighton, F., and Shamoon, T., Secure Spread Spectrum Watermarking for Multimedia, *IEEE Transactions on Image Processing,* vol. 6, no. 12, 1997.

[40] Hartung, H., Su, J. K., and Girod, B., Spread Spectrum Watermarking: Malicious Attacks and Counterattacks, *Proceedings of SPIE, Security and Watermarking of Multimedia Contents,* vol. 3657, pp. 147–158, 1999.

[41] Liu, H., *Signal Processing Applications in CDMA Communications,* Artech House, Norwood, MA, 2000.

[42] Barni, M., Bartolini, F., Cappellini, V., and Piva, A., A DCT-Domain System for Robust Image Watermarking, *Signal Processing,* vol. 66, no. 3, pp. 357–372, 1998.

[43] Swanson, M. D., Zhu, B., and Tewfik, A. H., Transparent Robust Image Watermarking Transform, *Proceedings of IEEE International Conference on Image Processing,* 1996, pp. 211–214.

chapter 6

Error concealment in digital video

Francesco G. B. De Natale

Contents

0-8493-1526-3/2004/$0.00+$1.50
© 2004 by CRC Press LLC

6.1 Introduction

Video streaming is strongly affected by errors and losses introduced by transmission systems. Although traditional transmission systems are quite robust to noise and interference, the problem of error control is becoming a crucial one due to the increasing use of mobile channels and packet-switched networks in distributed multimedia services. Such channels are characterized by high error rate, error bursts, packet losses, and strong temporal variability.

The impact of errors on coded multimedia streams is amplified by the use of source compression techniques, which are required for efficient multimedia data delivery. As a matter of fact, current video coding standards rely on vulnerable codes, subject to data loss and synchronization problems. Video error control procedures are targeted at removing or at least reducing the impact of errors on the rendition of the decoded video stream. A comprehensive review on the subject can be found in [1].

These techniques can be roughly subdivided into two families:

1. Channel-based techniques, in which the goal is to protect the message before transmitting it. These include:
   - Forward error correction codes (FECs), which are very easy to implement, independent of the specific data flow (according to the Shannon theory), and applicable in any situation. They cannot guarantee an error-free stream, and they produce a certain amount of overhead.
   - Retransmission protocols (ARQ), which are the best solution to avoid transmission errors but are inappropriate for many applications. For instance, they cannot be used in broadcast or multicast services or where the uplink channel is missing. Furthermore, they are ineffective whenever latency is a problem and can worsen the situation in error-prone channels, where they can generate huge retransmission and congestion problems.
2. Source-based techniques, which exploit the characteristics of the data source to improve the reliability of the transmission. These include:
   - Error-resilient video coding, in which the characteristics of the compression technique and of the packing format are designed to make the code intrinsically more robust to errors. In this case, the problem is handled *a priori* by transmitting a compressed video stream less vulnerable to loss and synchronization problems.
   - Error concealment, in which the video stream is postprocessed in order to mask possible errors from a perceptive viewpoint. In this case the problem is handled *a posteriori*, i.e., at the decoder, without necessarily requiring an interaction with the encoder.

In the last few years, there have been several efforts to combine these approaches with the aim of achieving a higher degree of error resilience. A significant example in this respect is represented by *joint source-channel coding* techniques, where error-correction codes are adaptively applied to the data stream, taking into account the relative importance and priority of the source information. To attain this goal, several methods have been proposed, including video layering and hierarchical coding, use of feedback messages, reversible codes, and so on. Also, in the case of error concealment, techniques have been studied that exploit a particular design of the source code such as layering and interleaving to facilitate the following concealment.

In the following sections, after a short review of the effects of errors on the decoded video stream, we will focus on the error concealment problem. We will first classify the different approaches to the problem, and then we will analyze the most significant methodologies produced in this field by the scientific community in the last years. Although we will focus mainly on the problem of concealing errors in video, several approaches presented in this chapter apply to still picture and binary shapes, as well.

6.2   Effect of errors in video coding

Block transform coding, the discrete cosine transform (DCT) in particular, has become the most popular approach to image and video coding as well as the core of the current international standards for storage and transmission of visual data, such as ISO JPEG and MPEG suites and ITU H.26x [2–6]. The common structure of the encoder includes a decomposition of the image in bidimensional blocks, the independent application of the transform to each block, and the quantization of the resulting coefficients to achieve the desired compression. The quantized coefficients are then arranged according to a standard-specific format and passed to an entropy coder to exploit the residual correlation. Variable length codes (VLCs) such as Huffman or arithmetic coding are usually adopted to this purpose to achieve best performance. Additionally, in video sequences the temporal correlation among consecutive frames is exploited by means of predictive techniques such as motion estimation and compensation.

The above operations allow for strongly reducing the amount of data associated with the visual object but make the relevant stream extremely vulnerable to possible errors introduced by an unreliable communication channel. As a matter of fact, a packet loss or even a single bit error on a VLC-encoded stream produces desynchronization of the decoder and often the impossibility of reconstructing the following codewords until the next resynchronization code is found. The use of predictive techniques worsens the problem, for it produces a temporal propagation of the error. The effects on the visual quality are often catastrophic and have to be correctly managed to ensure guaranteed quality of service (QoS).

Different channels and transmission protocols are usually associated with different error statistics. For instance, radio links used in mobile communications suffer from elevated BER with possible error bursts, as well as strong temporal variability; wired links provide a more stable bandwidth with lower error rates, which become quite negligible in optical networks. From the viewpoint of the transmission protocol, packet networks are subject to loss of entire packets due to timing and errors in packet headers, whereas in fast packet switching (e.g., cell relay technologies), packet losses are mainly caused by cell discarding at the network nodes due to congestion problems.

Bit errors can corrupt the stream at unpredictable points, causing a variety of faults. If the error occurs in the area containing video data (coded coefficients, motion vectors, etc.) or other local information, it propagates in any case until the next synchronization code is found, due to the use of VLCs. Depending on the point where the error occurred and the distance between successive synchronization codes (usually big enough to avoid excessive overhead), the error can produce long chains of lost image blocks. If the damaged frame is used as a reference for motion compensating a neighboring frame, the loss may also compromise the reconstruction of the whole spatio-temporal context of the lost area, due to the fact that erroneous information is used in the compensation. This problem is particularly troublesome in intracoded pictures. Figure 6.1 shows the effects of spatial and temporal propagation on a set of consecutive frames of a video sequence.

The use of reversible VLCs (RVLCs) can reduce this problem at the price of a higher complexity of the decoder and lower efficiency of the code itself [7]. In fact, RVLCs can be decoded in either direction, thus allowing recovery of a part of the lost sequence moving from the next synchronization point backwards to the error. Another possibility to prevent the loss of large image areas is the use of interleaving. In block interleaving, successive frame blocks (or macroblocks) are rearranged in order to be transported by different data

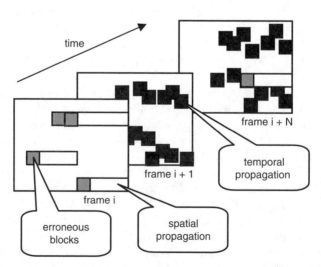

Figure 6.1 Spatial and temporal error propagation in a video sequence.

packets. In this way, a packet loss generates sparse block losses, which are much easier to conceal.

The effect of a bit error on a more sensitive area of the code (e.g., a header or a synchronization code) can be even worse, causing a complete loss of large image/video parts, depending on the level of the header. Sometimes this problem is managed by introducing redundancy or higher protection on those sections, again at the price of an increased overhead.

In most recent video coding standards, making use of complex scene descriptions and video objects (in particular, MPEG-4), the error can also partially destroy the object description, making it impossible to decode an entire video object. This is quite a new problem from the research standpoint, although some efforts have already been spent in this domain of investigation [8–10].

Finally, the error can happen in a vital part of the transport protocol (e.g., a packet header), resulting in the complete loss of the payload that, in turn, may cause the loss of large portions of the stream. This is generally associated with severe losses in the image and complete desynchronization of the decoder, which could also fail to calculate the extension of the loss and the position to which successive received data refer.

A further problem in streaming is due to packet delay. In this case no error occurs, but data packets can be delayed due to network problems (e.g., a congestion in a UDP transport node) and can eventually be overcome by the following packets that were routed via a different path. To solve this problem, complex packet buffering and reordering procedures are needed, which can fail when an excessive delay causes the rejection of the packet. More uncommon, packet intrusion (i.e., the insertion of a packet belonging to a different connection due to errors in addressing) can also have catastrophic consequences.

A complete description of these problems is behind the scope of this dissertation. In [11] an accurate model of an error-prone channel and the relevant impact on video transmission performance can be found.

6.3 Classification of video concealment approaches

Among other tools to achieve error resilience, concealment techniques provide several advantages. First of all, concealment can be implemented as a feature of the decoder, not requiring any modification of the encoder. It can imply different levels of complexity according to the desired result and to the computation and processing time requirements. It does not imply any additional data to be transmitted, thus avoiding overhead and congestion problems. Furthermore, many concealment approaches are standard independent and therefore intrinsically compliant with different standards and their progresses.

The general approach consists in masking the effects of errors to the observer by identifying a damaged area and substituting it with a patch able to deceive the human visual system and make it difficult to perceive the error. This is usually possible thanks to the inherent spatio-temporal redundancy of the video data, which allows to interpolate, in a more or less sophisticated way, the missing data. Interpolated data do not necessarily have to provide a high fidelity reconstruction of the missing part, just a plausible restoration of it. In this sense, it is not easy to evaluate the concealment performance, which is not necessarily related to the peak signal to noise ratio (PSNR) with respect to the original undamaged data.

The common basis of all the approaches so far proposed is in the attempt to interpolate or extrapolate the lost information by means of the correlated data available in the surrounding areas. The main differences lie in the type of correlation taken into consideration and in the way the algorithm exploits it. A first classification can be done between:

- Methods involving both the encoder and decoder
- Methods involving only the decoder

The former require the availability of a return channel, which is used to provide a feedback to the encoder. The latter work *a posteriori*, without requiring any interaction with the encoding and transmission system. The first approach usually shows better performance but lacks flexibility, for it must fit a specific encoding and transmission system. Moreover, it is not suitable for broadcast/multicast applications and pre-encoded video.

A further cataloguing should consider the type of correlation exploited by the algorithm. According to this, four different categories can be identified:

- Techniques based on spatial correlation
- Techniques based on spectral correlation
- Techniques based on temporal correlation
- Hybrid or composite techniques

The first category includes several approaches aimed at interpolating the missing data from the neighboring spatial information available within the same frame. It suits both video and still picture and works correctly also in the presence of fast movement, scene changes, and intracoded frames. The main disadvantage of these techniques lies in the fact that spatial correlation is usually lower than temporal correlation. This prevents the possibility of achieving a good reconstruction when the lost area has a large extension or contains small details or textures, in which case the concealed area may show visible blurring and perceivable temporal discrepancies.

Although often treated as a separate category, frequency-based techniques can be considered a subset of space-based approaches, where the spatial information is directly exploited in the transformed domain. These techniques have the advantage of better reconstructing the frequency components present in the data while reducing the number of significant variables in the problem (due to the energy compaction principle). The major drawback consists in the need to interact with the decoding process, which reduces the flexibility of the technique.

Temporal concealment works only for video sequences. It exploits the temporal correlation among successive frames by extrapolating the missing information from previous or past available data. It performs very well in the presence of slowly moving images or to reconstruct a static background where temporal correlation is particularly high. However, it produces unreliable results in fast moving areas, scene changes, or where occlusion problems are present.

Finally, several hybrid techniques have been recently proposed that aim at combining two or more of the above approaches to achieve better performance over a wide range of situations. In this framework, techniques integrating spatial and temporal concealment can obtain almost perfect restoration of lost blocks while requiring higher computation and storage requirements that make them often unsuitable for real-time applications.

In our survey we will focus mainly on postprocessing-based approaches, which are the subject of the next section, and in Section 6.5 the techniques involving some interaction with the encoder will be briefly discussed.

6.4   Concealment based on postprocessing

In this section, we will review the main approaches based on postprocessing at the decoder, without the need of interacting with the encoding and transport parts of the system. We will treat, in different subsections, techniques based on spatial correlation (in either pixel or transform domain), temporal correlation, and hybrid.

6.4.1   Techniques based on spatial correlation

Space based concealment techniques exploit the intraframe correlation present in natural images to interpolate the damaged area. The first investigations in this domain were targeted to the recovery of errors in still pictures, but they became the basis for several developments in video concealment. We will consider five different classes of methods, approaching the problem with different philosophies: interpolation techniques, structure-based approaches, prediction-based approaches, projections onto convex set, and statistical methods. For each class we will highlight a selection of techniques proposed in recent years.

6.4.1.1   Interpolation Techniques

Interpolators for error concealment are based on the observation that natural images are characterized by a low graylevel variation, so that the replacement of the damaged pixels with a smooth surface, gracefully linked with the neighboring area, would provide a sufficiently accurate error masking.

In [12], Wang et al. proposed a technique called *maximally smooth interpolation*. Here the objective was to reconstruct the missing DCT coefficients in the damaged block (or a subpart of them) by generating a set of transform coefficients that guarantee the maximum possible smoothness at the boundaries of the lost area. To this aim, an energy function is minimized, taking into account the variation across the boundary of the lost block and extending the smoothness constraint to the internal points, as shown in Figure 6.2.

A limitation of this approach that is common to all DCT-based interpolation techniques lies in the fact that it is not suited for concealment of intercoded frames, where the correlation among residual coefficients is very low. Another drawback of maximally smooth interpolation lies in the blurring effect, particularly visible in restored blocks crossed by edges. An improvement in this respect is presented by the same authors in [13], where the first-order derivative present in the cost function is substituted with a second-order derivative. The second-order derivative is estimated in the discrete domain by combining a quadratic variation measure and a Laplacian-based smoothness measure, thus achieving a slow variation along edges while reducing the blurring across them.

In [14], Park et al. proposed a slightly different conceptual approach, which leads to a very efficient implementation of a DCT coefficient recovery-based interpolation. In their algorithm, a quadratic object function is defined that introduces a smoothness constraint only at the borders of the block to be recovered. The object function is then maximized by solving a differential system, in which the vector of unknowns is limited to the dominant low-frequency DCT coefficients. This allows for attainment of a suboptimal solution with a very limited number of equations, thus reducing the computation in comparison with competing techniques.

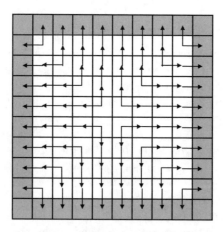

Figure 6.2 Border constraints imposed by the maximally smooth interpolation. The arrows show the pixel pairs, the variation of which is taken into account in the energy function, whereas the shaded pixels represent the external boundary (available at the decoder). The number of constraints can be reduced if only part of the DCT coefficients are lost.

An interesting idea to include directional information in the interpolation process was proposed by Kwok and Sun [15]. They classify the blocks surrounding the lost one in eight directional classes according to their estimated spatial gradient, and they extrapolate them accordingly in the damaged area. The individual directional extrapolations are then mixed together to achieve the recovered block by using a set of heuristic rules.

An attempt to further refine this idea can be found in [16], where a better knowledge of the direction of the main edges crossing the lost area is achieved by a Hough transform. The transform is applied on a 48 × 48 pixel window centered at the lost block in order to detect the presence of significant rectilinear lines. If any such line is found, a directional interpolation is applied through filtering.

6.4.1.2 Structure-based concealment

In order to improve the rendering of high spatial frequencies, and in particular contours, it is necessary to take into account more explicitly the geometric structures present in the image. Several techniques have been proposed in this framework.

In [17], edges are extracted from the region surrounding the lost blocks and are integrated with the relevant graylevel information to build a sketch. The sketch is then extrapolated in the area to be concealed and used as a basis for the reconstruction of low-frequency and textured areas, which are then recovered by means of smooth interpolation and patch replication functions. The most complex part of this scheme consists in determining the correct matching among the sketch fragments to be joined; this operation is carried out by an optimization procedure based on a composite cost function. As compared to smooth interpolation, sketch-based concealment provides a quite natural appearance also in the presence of edged areas, where contours are adequately interpolated. Nevertheless, some problems may arise when the sketch-matching procedure fails, thus generating a wrong coupling of the contour segments.

Zeng and Liu exploited the geometric structures present in the image to recover the damaged image portion [18]. In their work, a computationally efficient spatial directional interpolation is proposed with application to both error concealment and voluntary block dropping in video transmission over very-low-bandwidth channels. First, the presence and orientation of transition points at the boundary of the lost block is detected by applying a simple thresholding algorithm along a two-pixel-wide external border. According to the number of transition pairs, the block is classified as smooth (no transitions) or edged (with 1, 2, 3, or more transition pairs). After that, the geometric structure is inferred by connecting the transition pairs according to the relevant orientations, and a directional interpolation is applied to the recovered block structure. As in [17], the most challenging problem is to correctly infer the structure; for this reason, a set of specific rules is defined for each of the above five classes.

6.4.1.3   Projections onto convex sets

Projections onto convex sets (POCS) were originally proposed by Sezan and Tekalp [19] with application to image restoration. On this basis, Sun and Kwok proposed in [20] an interesting approach to image concealment. They consider a spatial context of the lost block made up of the eight connected blocks (called large block), which is Sobel filtered in order to detect possible edges. Based on the extracted information, the block is classified as *monotone* or *edged*, and in the last case it is further classified according to the edge direction, quantized in eight equally spaced angles. At the same time, the large block is Fourier transformed for successive processing.

After that, two different convex projections are applied. The first operator is based on an adaptive filtering of the large block according to the classification. This operation is performed in the Fourier domain imposing convexity constraints on the coefficients. The second operator maps the filtered block in the spatial domain by imposing two convex constraints: the values at the known blocks have to remain unchanged, and the values are clipped in [0–255] to fit the original 8-bit representation. This filtering and copy-back process is iterated until the variation in the central block becomes negligible; usually a few iterations are sufficient, thus ensuring a reasonable computation. In Figure 6.3 a conceptual scheme of this POCS restoration process is shown.

In [21] a quite sophisticated strategy is proposed, using object models and iterative replenishment based on the POCS theory. In this technique, appearance models are associated to specific objects in the scene, and they are used to restore the lost area. This approach is particularly suitable in the framework of MPEG-4 standard, due to its object-based nature, but can be extended to any other coding technique provided that an effective modeling of scene objects is attained. The technique proposed for model building is called mixture of principal components (MPC) and is formulated as an extension of the principal components analysis (PCA) and able to catch the large variations present in video objects.

Once the model is available, the concealment is performed in two steps: (i) projection of the region of interest into the model to achieve the reconstruction and (ii) replacement of missing data in the region of interest using the reconstruction. It is demonstrated that both these steps can be regarded as projections onto convex sets, thus allowing an iterative solution; to this pur-

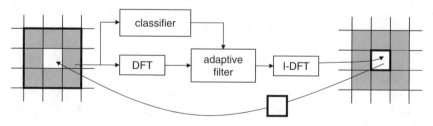

Figure 6.3 Scheme of the POCS adaptive restoration process.

pose, steps (i) and (ii) can be iteratively repeated to obtain a better result. Tests performed mainly on face sequences show a significant improvement in PSNR performance over zero-motion and motion-compensated replenishment.

6.4.1.4   *Prediction-based approaches*

Another possible approach to spatial concealment consists in generating an estimate of the lost information, using the available data as an input to a prediction scheme. The problem of predicting a lost block or block sequence is a complex one, due to the high dimensionality of the input and output data. Even considering a single 8×8 block, the estimation of 64 parameters with an 8-bit range of variation can be extremely troublesome for every prediction algorithm.

In [22] a genetic algorithm (GA) is used to this purpose: the 8-bit gray-level values at the 64 pixels of the lost block are encoded in a 512-bit chromosome, which has to evolve to the optimum solution with the objective of minimizing a block fitness function. The fitness can take into account both spatial and temporal parameters, linearly combined with appropriate weights. The method can be adapted to static images or intracoded frames by exploiting only five spatial components, defined as performance parameters: the average intersample difference within the block (α), the average intersample difference across block boundaries (β), the average α difference (γ), the average mean difference (δ), and the average variance difference (λ). As far as the initialization of the GA is concerned, an initial population is set up including randomly generated patterns and user-specified bit strings defined on the basis of the spatio-temporal context of the lost block.

In [23] the problem is faced directly in the transform domain by using a neural predictor. To this purpose, a three-layer Perceptron is trained to reproduce the DCT coefficients of an isolated missing block, given in input a set of coefficients taken from the surrounding available blocks. In order to reduce the complexity of the problem, only a subset of the coefficients is selected in input and output blocks. In output configuration, only low-frequency components are chosen, mainly in the horizontal and vertical directions, thus ensuring the smoothness of the reconstruction and an easier convergence of the solution. As far as the input is concerned, the spatial frequencies are selected according to the relative position of the blocks (see Figure 6.4). The most difficult aspect of this method is in the definition of the training procedure. As a matter of fact, a wrong selection of the training set can prevent effective learning. The authors propose the use of synthetically generated patterns as well as the accurate selection of samples from real images.

6.4.1.5   *Statistical approaches*

Statistical approaches are based on the modeling of the lost sequence as a random process and of the known information as a conditioning element. The recovery process can be implemented as a maximum likelihood (ML)

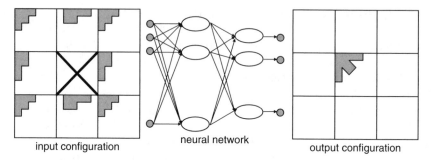

input configuration   neural network   output configuration

Figure 6.4 A DCT-based neural prediction scheme: the shaded positions in the input/ output configurations correspond to the coefficients selected as input and output, respectively; the central block (marked with an × in the input configuration) is the lost one; the circles in the neural architecture represent the Perceptron weights that are adjusted during the learning phase.

or maximum *a posteriori* (MAP) estimation applied to the conditional probability of the lost data.

In [24] a Bayesian approach is proposed, in which the MPEG video sequence is modeled as a Markov Random Field (MRF) and the lost blocks are calculated by maximizing the *a posteriori* conditional probability. Let x be a decompressed MPEG sequence and y its damaged version. The goal is to achieve the optimum estimate of x given y. According to the MAP scheme, the optimum estimate (which incidentally coincides with the minimum error probability) can be computed by maximizing the conditional probability $f(x \mid y)$, where x is modeled as an MRF, namely

$$f_x(x) = \frac{1}{Z} \exp\left(-\sum_{c \in C} V_c(x) \right)$$

where Z is a normalizing factor, c is the so-called *clique* (local group of points), C is the set of all cliques, and $V_c(x)$ is a quadratic cost function. The optimum estimate $\hat{x}$ is then given by

$$\hat{x} = \arg \min_{x \mid y = Dx} \left(\sum_{c \in C} V_c(x) \right)$$

where D is a matrix accounting for the loss of information between x and y. An analogous procedure can be applied to recover the lost motion vectors in intercoded frames.

Although this model is quite elegant from the theoretical viewpoint, a major problem with MRFs is in their intrinsic hypothesis on the stationary nature of the source. This assumption can be considered reasonable in smooth image areas, yet it does not hold in the presence of fast transitions, like edge points. In [25] Li and Orchard try to solve this problem by moving from a blockwise estimation to a sequential pixelwise approach. To this aim, they rewrite the conditional probability density function in a series of pixelwise conditional PDFs, which are sequentially solved by using the previous results as additional conditional elements. The individual estimations are performed by using an orientation adaptive interpolation based on the computation of the local covariance matrix. During this operation, no distinction is made among successfully received data and previously recovered ones. This choice simplifies the model but can potentially lead to severe error propagation. This problem is faced by performing the sequential scanning along different directions and then linearly merging the results. The possibility of adopting nonraster scanning order is also considered.

6.4.2   Techniques based on temporal correlation

Time is associated to a large correlation in moving images. Except for the case of scene changes or chaotic motion, successive frames of a video sequence are very similar to each other, thus providing a precious source of information whenever a restoration process is to be applied. A simple and effective approach to temporal concealment consists in replacing the lost block with its motion compensation. This operation, known as motion-compensated temporal prediction, is possible in general if the relevant motion vector (MV) is available; this is the reason why motion field information is usually transmitted as high-priority protected information in the video stream and sometimes transmitted also for intracoded blocks.

However, when the block information is completely unavailable, i.e., the motion information and possibly also the block coding mode are lost, the problem becomes somewhat more complex. A possible strategy consists in reconstructing *a posteriori* the provenance of the damaged area with respect to a previous frame. The lost block can be then approximated by a displaced and possibly modified version of the corresponding block in the previous frame (see Figure 6.5). This operation is referred to as motion field recovery and can be performed by exploiting the spatio-temporal smoothness of the video sequence. Several approaches have been proposed to achieve this goal. We roughly subdivide them into two classes: methods that use only motion information and methods that also incorporate spatial information for motion prediction.

6.4.2.1   Approaches based on motion field interpolation

Very simple approaches to motion field recovery consist of interpolating or extrapolating (predicting) the motion information of the relevant area in a previously received frame. To this purpose, several possibilities were explored, including [26]:

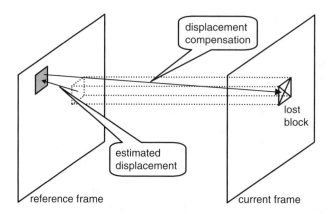

Figure 6.5 Conceptual scheme of error concealment by motion field recovery.

- Zero motion: the motion vector is supposed to be null, so that the lost block is replaced with the block at the corresponding position in the reference frame. It works well in static areas or background but poses severe problems in moving areas.
- Temporal replacement: the motion vector is substituted by the MV at the same position in the reference frame or by a combination of the surrounding ones.
- Spatial mean: the motion vector of the current block is estimated as the average of the motion vectors of the spatially neighboring blocks
- Spatial median: the motion vector of the current block is estimated as the median of the motion vectors of the spatially neighboring blocks

The above can also be combined on a block or pixel basis to achieve better adaptation. As an example, Al-Mualla et al. proposed in [27] two motion field recovery approaches that are individually applied to each pixel of the lost block. It is shown that this can be advantageous when the motion inside the block is inconsistent (e.g., it contains parts of objects moving in different directions) or when the global motion estimation is not satisfactory. In the first technique, called motion field interpolation (MFI), the motion of each pixel is estimated as the weighted sum of the MVs at the neighboring blocks. In the second, called multihypothesis motion compensation (MHMC), the lost block is compensated using a weighted average of two motion-compensated predictions, achieved with MFI and BMA (see the next subsection), respectively.

In [28], an alternative solution is presented that solves the optical flow (OF) equation to gain a better approximation of the object motion. To this end, the unavailable motion vector is calculated as the average of OFs within a region surrounding the MV estimated block.

Shanableh and Ghanbari [29] demonstrated that interpolated frames are particularly important for many applications such as transcoding and error resilience. This is due to the fact that B-frames carry information on both past and future anchor frames, making it possible to backtrack the best-matched location in the future anchor frame to the best-matched location in the previous one. In their work, they propose a method for concealment of lost anchor frames, accounting for the fact that errors in other frames do not propagate and are therefore less critical. If any bidirectionally predicted frame is present among two damaged anchor frames, a lost macroblock in an anchor frame can be concealed as shown in Figure 6.6. A part of the block (in white) is directly replaced with the corresponding matching area in the other anchor frame. The remaining part (in gray), which remains uncovered due to macroblock misalignment, is replaced by the area pointed to by the sum of the two motion vectors, namely:

$$V_P^{fwd} = V_B^{fwd} - V_B^{bwd}.$$

A problem arises when bidirectional prediction is not used in the lost area in any B-picture in between the two anchors. In this case, the above technique cannot be applied, and more complex strategies based on selection and combination of the available MVs are proposed.

In [30] the same authors extended this method by applying the restoration process to all inter- and intraframes. Moreover, they propose to impose some constraints in the motion field computation in order to ease the recovery process. Different approaches are presented to this end, which may or may not maintain the compliance to MPEG syntax, and the impact on the quality and compression of the resulting stream is analyzed in detail.

6.4.2.2 *Approaches using spatio-temporal information for motion recovery*

The idea in this case is to jointly exploit the spatio-temporal correlation present in the data to estimate the area in the reference frame that better matches the lost area in the current one. A very good intuition in this direction is due to Lam, Reibman, and Liu [31]. They proposed to estimate, *a posteriori*, the motion vector of the lost block by performing a block-matching procedure based on

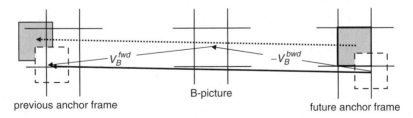

previous anchor frame   future anchor frame

Figure 6.6 Backward tracking of MVs: the lost macroblock in the future anchor is compensated partially by the corresponding block in the past anchor (white area, bold arrow), and partially by the block pointed to by the sum of the forward and backward MVs at the B-picture (gray area, dotted arrow).

the border information. They called this technique boundary matching algorithm (BMA). In BMA, the MV is selected among a set of candidates by minimizing the quadratic luminance difference between the external boundary of the lost block and the corresponding pixels at the displaced position in the reference frame. Although this operation can be theoretically performed as a full search (see [32] for a proposal in this sense), the authors speed up the computation by defining a set of candidates including the MV of the same MB at the previous frame and those of the neighboring MBs, as well as their average and zero motion vector.

Feng and Mehrpour extended this technique by including a border analysis aimed at identifying possible edges present in the lost area and modifying accordingly the boundary difference measure [33]. Another attractive variant was investigated by Lee et al. [34], who extended the boundary smoothness constraint of BMA to the successive P and B frames in the sequence. They showed how multiframe BMA could provide better rendering of the recovered area, in particular when horizontal edges are present.

In [35] a further improvement is proposed, based on block overlapping. Here, after finding the best-matching MV by a BMA-like procedure, the block is subdivided into four sub-blocks, and each sub-block is synthesized as a weighted average of three predictions: the block pointed by the MV calculated by BMA, and the two blocks pointed by the MVs of the horizontal and vertical neighboring blocks. The weights are set similarly to overlapped motion compensation in H.263.

6.4.3   Hybrid techniques

Hybrid concealment makes use of both spatial and temporal correlation in an integrated way to improve the reconstruction quality. Due to the higher possibilities offered by such an approach, several studies have been carried out in this direction in the last few years. We will consider three different strategies, and we analyze, in the following subsections, a selection of the proposed schemes for each of them.

6.4.3.1   Spatio-temporal replacement

Spatio-temporal replacement consists in reconstructing the damaged area by using the relevant information gathered from the neighboring area in the same frame and/or by the matching area in the neighboring frames. The simpler way to implement this function is to define a set of candidates among the adjacent blocks in a space and time context, and to select the concealed block as one of the candidates or a combination of them. Chu and Leou proposed in [36] a technique in which the set of candidates (SC) includes the received and concealed neighboring blocks of the lost one, their spatial average and median, the motion-compensated blocks in the previous frame pointed to by the MVs of the spatially neighboring blocks, the motion-compensated blocks in the previous frame pointed to by the average and median of the MVs of the spatially neighboring blocks, and their spatial average and median. The selection is then performed by choosing the candidate that minimizes a composite cost function, taking into account the smoothness of the transition across the block boundary, the difference between the spatial statistics of the replaced block and the neighboring ones, and the temporal compatibility of the replaced block with the past one at the same position.

In [37] the adaptation is based on local measures of temporal and spatial activity as well as on the block type. Blocks for which motion information is available are temporally concealed, while intracoded blocks are classified according to their activity and concealed accordingly. Where a high spatial activity is detected, motion-compensated temporal concealment is used; where temporal activity prevails, spatial interpolation is preferred.

Also, Tsekeridou and Pitas [38] proposed to alternate a spatial and temporal approach according to the frame type. A spatial method called *split-match* (SM) is applied to the first I-frame, where no temporal information is available, whereas a *forward–backward block-matching* (FB-BM) approach is adopted for the other frames. In SM, the block is subdivided according to the similarity of its texture with neighboring blocks. Two possible similarity measures are proposed, one based on the MAD computation and the other based on low-order statistics. Once the split is performed, each subpart of the block to be concealed is approximated based on the best-matching available texture. Three alternative approximation methods are used: copying along the best-match direction, anisotropic diffusion (based on the Perona-Malik model [39]), and bilinear interpolation of the border pixels. As for FB-BM, it is a temporal concealment technique based on block-matching principles where the major innovation consists in the fact that only the neighbor leading to the best possible concealment is used.

Lee et al. showed that a multiframe analysis is necessary to reduce the global distortion caused by an error in terms of both local damage and propagation to the following frames [40]. Their method is based on the so-called *multiframe recovery principle*, which explicitly analyzes the error propagation of the lost block and minimizes it across multiple frames. Figure 6.7 shows how this objective is achieved: the MBs in the future frames that include parts of the lost block after compensation are tracked, and the borders of the damaged areas are identified. In the subsequent processing, smoothness constraints are extended to the borders of the following predicted frames. In the spatial case, a kind of maximally smooth reconstruction is applied by taking into account the extended boundary, whereas in the temporal domain the proposed algorithm estimates the optimum recovered MV as the one that minimizes an extended boundary-matching criterion.

6.4.3.2 *Error concealment using motion and spatial deformation*

A major problem in temporal replacement techniques lies in the impossibility of taking into account complex motions such as rotation, scaling, and deformation. This leads to annoying blocking artifacts and edge mismatch among the concealed block and the surrounding region.

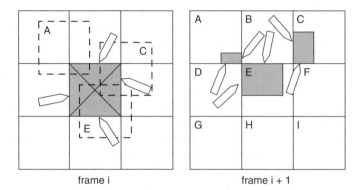

frame i   frame i + 1

Figure 6.7 Multiframe error concealment: the lost MB in *i*th frame (shaded block) propagates due to motion compensation in MBs A, C, and E in (*i*+1)th frame. Consequently, multiframe boundary smoothness constraints should be applied at the borders marked by the arrows.

Some approaches have been proposed to correct this problem. Lee et al. [41] introduced an affine transformation by dividing the lost block into triangular patches and computing the relevant affine transformation. The transformation parameters are computed by matching the vertices of each patch with the best-fitting points at the reference frame.

In [42] an affine transformation is used to improve the approximation obtained by boundary-matching algorithms. In this case, after finding the best-matching block according to a side match criterion, the patch is further modified by a warping process in order to match the contour points at the block boundaries. To this end, a set of control points is identified, corresponding to the position of edges in inner and outer boundaries of the concealed block, and the best-fit correspondences are found between internal–external control points. Based on the displacement of such control point pairs, a set of affine transformations is applied to warp the patch and achieve the final approximation (see Figure 6.8). It is shown that this approach can strongly reduce the border effects in the presence of fast or complex motion, making temporal concealment suitable also in the presence of high temporal activity.

6.4.3.3 *Statistical approaches*

Statistical approaches analyzed in Section 6.4.1.5 can benefit from a joint exploitation of spatial and temporal correlation. Shirani et al. [43] proposed a two-stage concealment that uses MRFs to refine the result achieved by temporal replacement. In their method, an initial estimate is achieved by a temporal analysis. To this purpose, MV information is used when available; otherwise, an interpolation algorithm is applied based on the maximization of the continuity inside the block and across its boundaries. For intracoded blocks, the initial estimate is set to zero. In the second stage, a MAP estimator

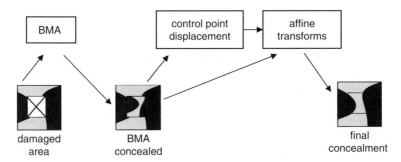

BMA — **control point displacement** → **affine transforms**

damaged area — BMA concealed — final concealment

Figure 6.8 Block-warping procedure by affine transformations in a rotational motion: the lost block is first substituted by BM patch and then deformed in order to fit the inner–outer edges.

is applied to a GMRF model of the image to improve the image reconstruction quality. The model is made adaptive by using a set of weights that are locally tailored to the likelihood of edges at each pair of points. The result of this second step is to maximize the smoothness along edges while maintaining sharp discontinuities across edges.

Zhang and Ma further improved the above concealment technique by including both spatial and temporal information in the statistical model [44]. They showed how this integrated model can effectively reduce the two major problems of traditional MRF models, i.e., the convergence to local minima and the blurring of contours. In their proposal, the spatial and temporal contextual features of a video sequence are individually modeled as multi-scale MRFs, leading to the so-called *dual multiscale MRF* (DM-MRF). The *dual* nature of the approach concerns the fact that both the image pixel field and the motion field can be modeled as MMRFs. The term *multiscale* refers to the fact that the estimation is performed on a sequence of nested subspaces of the original configuration space by applying a coarse-to-fine optimization process. On this basis, the lost information is approximated by a MAP optimum estimator, where the optimization function contains terms relevant to both space and time.

6.5 Concealment vs. source/channel coding

In this section, we will briefly analyze the concealment approaches that involve in some way the encoding process. We will subdivide the presentation into three subsections. The first describes techniques in which no direct interaction is needed between encoder and decoder but some structuring of the code is exploited, such as layering, overlapping, or interleaving. The second subsection will focus on the approaches that exploit channel coding. The third subsection will present methods that directly imply an exchange of data between encoder and decoder.

6.5.1   Concealment in structured video

Video error concealment algorithms can greatly benefit from particular structures or constraints applied at the encoder level or during transmission. These constraints can be compliant with the standards or may require some modifications. In our review we will mainly focus on standard-compliant techniques.

A very interesting feature to achieve error resilience is video layering. Layering consists in subdividing the video data into different streams (usually two streams: base and enhancement) that are transmitted with a different level of protection/priority. Usually, the base layer contains vital information on the video stream and sufficient data to provide a coarse rendering of the decoded sequence. The enhancement layer provides additional information for quality refinement. In [45], Ghanbari and Seferidis demonstrated that video layering can be also very advantageous for error concealment. In their work they exploited the availability of the motion field at the decoder, transmitted within the base layer over a guaranteed connection, to improve the performance of motion-compensated temporal replenishment. This implies that the motion field is related only to the base layer. An important result of this study was to demonstrate that better interpolation quality can be achieved by performing the motion estimation on the original frames instead of the coded ones, despite not being optimum in terms of bit-rate reduction.

In [46], two new approaches are proposed for error recovery in layered video. The first, called switch-per-pixel error concealment (SPEC) is a pixel-wise adaptive restoration that adaptively switches between two sources of information: the adjacent pixels in the base layer of the current frame and the corresponding pixels in the enhancement layer of the previous one. The binary decision is taken according to the base layer residue and the enhancement layer error propagation history. The second approach, called recursive optimal per-pixel estimate (ROPE), acts at the encoder level, allowing standard compliant optimal decision of the prediction modes used in the enhancement layer. The two approaches can be complemented to achieve additional performance gain.

Block interleaving [47] is another attractive feature, particularly suited to packet video transmission that has a big influence on the concealment performance. Although not explicitly mentioned, it is an underlying hypothesis in many of the concealment schemes so far analyzed. Interleaving consists of changing the order of coded blocks in the transport stream of in order to avoid the loss of consecutive blocks. The main advantage is in the possibility of fully exploiting the spatial correlation among neighboring blocks, making it more difficult to have large lost areas. The main drawback consists in the buffering and delay requirements. In [47], Zhu et al. proposed to integrate interleaving with video layering and an adaptive space–frequency–time concealment technique in order to achieve reliable video streaming over networks characterized by very high packet loss ratios.

In the field of noncompliant techniques, some interesting approaches have been proposed using block overlapping. Overlapping consists of the superposition, with partial data replication, of consecutive image blocks in the spatial domain. It results in a slightly higher bit rate and sometimes in problems related to nonconventional block dimensions. For instance, in [48] a 9×9 block structure is proposed, with a 1-pixel superposition of each block with the four connected ones. In the case of a block loss, the overhead is repaid by a more efficient recovery that uses the common boundaries present at the undamaged neighboring blocks instead of the external boundary. A modification of the maximally smooth criterion is used to this purpose. The same authors proposed in [49] a POCS-based approach based on this overlapping structure.

6.5.2 Concealment using channel coding

Although FECs can be applied as a transparent feature of the transmission protocol, it can be interesting to understand how their application interacts with the global performance of the transmission system. In [50] a very useful investigation in this direction is presented, with application to layered video coding. The authors start with a thorough comparison among two transmission modes. In the first, FECs are applied to the base layer only, whereas in the second, FECs are equally applied to both base and enhancement layers. Passive (postprocessing-based) concealment by temporal extrapolation is used in either case. This investigation leads to the conclusion that in most cases of practical interest, the first option is very convenient.

Based on this assumption, the authors proposed a third hybrid option, aimed at further limiting the bandwidth expansion due to channel coding while ensuring a high-quality overall reconstruction. The idea behind this hybrid scheme is to concentrate the error protection (and therefore the overhead bits) in the areas that are more difficult to conceal with traditional interpolation/compensation algorithms. Since those areas are recognized to be the ones characterized by a high temporal activity, the degree of local motion of the video sequence is first analyzed, and a decision device is inserted into the coding and transmission scheme with the task of switching on and off the FEC according to the motion.

A more recent approach using a channel-coding-like technology to improve the video error correction capability is the use of steganography. Steganography consists of hiding some data in the visual information that are impossible to be perceived by a user but contain useful information for successive processing. Examples of common application of steganography are in the field of copyright protection and watermarking. In [51] Robie and Mersereau proposed this kind of approach, in which the information hidden in the bit stream allows an early resynchronization of the video stream as well as the recovery of the differential DC coefficients. The only price to pay is a small degradation of the undamaged video quality, with a very limited increase in computational complexity. In [52] Carli et al. defined an error control and concealment scheme working in a similar manner but using wavelets to achieve an embedded version of the data to be hidden in the stream.

6.5.3   Concealment requiring interaction with the encoder

As mentioned in the introduction, the availability of a return channel is impractical for many applications. Nevertheless, in applications in which the interaction between encoder and decoder is feasible, it can be quite advantageous. A comprehensive review of the applicability of feedback in video transmission and of the relevant techniques for error control can be found in [53], with particular application to mobile video communication.

Concerning the specific use of a feedback channel for video concealment, one of the earliest investigations is due to Wada [54]. In his work, he proposed to send an asynchronous feedback in the presence of a packet error without stopping either encoder or decoder activity or requiring retransmission. The effect of such signaling simply avoids the further use of damaged blocks by the encoder until a refresh of the relevant image area is transmitted. The encoder can act in two possible ways: (i) build a prediction tree identifying the propagation of the lost block and exclude all the relevant information from successive processing steps or (ii) reproduce at the encoder the same concealment adopted by the decoder in order to work on the same data available at the receiver. The second approach is clearly more efficient, but unfortunately it is not suitable for multipoint transmission. In either case the result is a complete removal of error propagation, accompanied by a slight increase of the bit rate due to imperfect compensation of the motion around the lost blocks or to the necessity of intramode transmission (according to the transmitter actions). More recently, Wang and Chang further worked on this type of approach, studying its application to MPEG video streaming [55].

Finally, in [56] the selection of the algorithm to be applied for the concealment of each damaged block is taken through a decision tree, which is transmitted by the encoder. The authors propose switching among a large set of possible concealment schemes (including several spatial, temporal, and frequency methods) under the guidance of the encoder, which generates a parallel stream of *a priori* information on the local characteristics of the missing block, of its neighbors, and of the whole frame. It is to be observed that such an approach is not standard compliant.

6.6   Conclusions

We have analyzed the effects of the transmission of coded video streams over error-prone channels, such as mobile and packet networks, as well as some of the most interesting approaches so far proposed to mask the relevant damages to a human observer. These approaches were classified according to the strategy and to the type of features used in the concealment process. Further, a selection of the most significant approaches was analyzed in detail.

Several related points were left untreated due to space constraints. Among them, the problem of early error detection and resynchronization of the decoder and the problem of the assessment of concealment results in terms of objective and perceptive visual quality are very significant and open research topics. Links to these aspects can be found in the list of references.

References

[1] Wang, Y. and Zhu, Q.-F., Error control and concealment for video communication: a review, *Proceedings of the IEEE*, 86, 974, 1998

[2] Pennebaker, W. and Mitchell, J., *JPEG still image data compression standard*, Norwell, MA: Kluwer, 1992

[3] Mitchell, J., Pennebaker, W., and Fogg, C.E., *MPEG video compression standard*, Norwell, MA: Kluwer, 1996

[4] Koenen, R., Overview of the MPEG-4 standard, ISO-IEC JTC1/SC29/WG11 N1730, 1997

[5] ITU-T Recommendation H.261, *Video codec for audiovisual services at px64 kbits*, 1993

[6] ITU-T Recommendation H.263, *Video coding for low bitrate communication*, 1997

[7] Wen, J. and Villasenor, J., Reversible variable length codes for efficient and robust image and video coding, *Proceedings of the IEEE Data Compression Conference*, 1998, 471

[8] Shirani, S., Erol, B., and Kossentini, F., A concealment method for shape information in MPEG-4 coded video sequences, *IEEE Transactions on Multimedia*, 2, 185, 2000

[9] Li, X., Katsaggelos, A.K., and Schuster, G.M., A recursive shape error concealment algorithm, *Proceedings of the IEEE International Conference on Image Processing*, 1, 2002, 177

[10] Shirani, S., Erol, B., and Kossentini, F., Error concealment for MPEG-4 video communication in an error prone environment, *Proceedings of the IEEE International Conference on Acoustics, Speech, and Signal Processing*, 6, 2000, 4

[11] Stuhlmuller, K., Farber, N., Link, M., and Girod, B., Analysis of video transmission over lossy channels, *IEEE Journal on Selected Areas in Communications*, 18, 1012, 2000

[12] Wang, Y., Zhu, Q.-F., and Shaw, L., Maximally smooth image recovery in transform coding, *IEEE Transactions on Communications*, 41, 1544, 1993

[13] Zhu, W., Wang, Y., and Zhu, Q.-F., Second-order derivative-based smoothness measure for error concealment in DCT-based codecs, *IEEE Transactions on Circuits and Systems for Video Technology*, 8, 713, 1998

[14] Park, J.W., Kim, J.W., and Lee, S.U., DCT coefficients recovery-based error concealment technique and its application to the MPEG-2 bit stream error, *IEEE Transactions on Circuits and Systems for Video Technology*, 7, 845, 1997

[15] Kwok, W. and Sun, H., Multi-directional interpolation for spatial error concealment, *IEEE Transactions on Consumer Electronics*, 39, 455, 1993

[16] Robie, D.L. and Mersereau, R.M., The use of Hough transforms in spatial error concealment, *Proceedings of the IEEE International Conference on Acoustics, Speech, and Signal Processing*, 6, 2000, 2131

[17] Atzori, L. and De Natale, F.G.B., Error concealment in video transmission over packet networks by a sketch-based approach, *Signal Processing: Image Communications*, 15, 57, 1999

[18] Zeng, W. and Liu, B., Geometric-structure-based error concealment with novel applications in block-based low-bit-rate coding, *IEEE Transactions on Circuits and Systems for Video Technology*, 9, 648, 1999

[19] Sezan, M.I. and Tekalp, M., Adaptive image restoration with artifact suppression using the theory on convex projections, *IEEE Transactions on Acoustic, Speech and Signal Processing*, 38, 181, 1990

[20] Sun, H. and Kwok, W., Concealment of damaged block transform coded images using projections onto convex sets, *IEEE Transactions on Image Processing*, 4, 470, 1995

[21] Turaga, D.S. and Tsuhan, C., Model-based error concealment for wireless video, *IEEE Transactions on Circuits and Systems for Video Technology*, 12, 483, 2002

[22] Shyu, H.-C. and Leou, J.-J., Detection and concealment of transmission errors in MPEG-2 images — a genetic algorithm approach, *IEEE Transactions on Circuits and Systems for Video Technology*, 9, 937, 1999

[23] De Natale, F.G.B., Perra, C., and Vernazza, G., DCT information recovery of erroneous image blocks by a neural predictor, *IEEE Journal on Selected Areas in Communications*, 18, 2000

[24] Salama, P., Shroff, N., and Delp, E.J., A Bayesian approach to error concealment in encoded video streams, *Proceedings of the IEEE International Conference on Image Processing*, 2, 1996, 49

[25] Li, X. and Orchard, M.T., Novel sequential error-concealment techniques using orientation adaptive interpolation, *IEEE Transactions on Circuits and Systems for Video Technology*, 12, 857, 2002

[26] Haskell, P. and Messerschmitt, D., Resynchronization of motion compensated video affected by ATM cell loss, *Proceedings of the IEEE International Conference of Acoustic, Speech and Signal Processing*, 3, 1992, 545

[27] Al-Mualla, M.E., Canagarajah, C.N., and Bull, D.R., Motion field interpolation for temporal error concealment, *IEEE Proceedings — Vision, Image and Signal Processing*, 147, 445, 2000

[28] Suh, J.-W. and Ho, Y.-S., Error concealment technique based on optical flow, *Electronics Letters*, 38, 1020, 2002

[29] Shanableh, T. and Ghanbari, M., The importance of the bi-directionally predicted pictures in video streaming, *IEEE Transactions on Circuits and Systems for Video Technology*, 11, 402, 2001

[30] Shanableh, T. and Ghanbari, M., Loss concealment using B-pictures motion information, *IEEE Transactions on Multimedia*, 5, 257, 2003

[31] Lam, W.-M., Reibman, A.R., and Liu, B., Recovery of lost or erroneously received motion vectors, *Proceedings of the IEEE International Conference on Acoustic, Speech and Signal Processing*, 5, 1993, 417

[32] Zhang, J., Arnold, J.F., and Frater, M.R., A cell-loss concealment technique for MPEG-2 coded video, *IEEE Transactions on Circuits and Systems for Video Technology*, 10, 659, 2000

[33] Feng, J., Lo, K.-T., and Mehrpour, H., Error concealment for MPEG video transmissions, *IEEE Transactions on Consumer Electronics*, 43, 183, 1997

[34] Lee, Y.-C., Altunbasak, Y., and Mersereau, R., A temporal error concealment method for MPEG coded video using a multiframe boundary matching algorithm, *Proceedings of the IEEE International Conference on Image Processing*, 1, 2001, 990

[35] Chen, M.-J., Chen, L.-G., and Weng R.-M., Error concealment of lost motion vectors with overlapped motion compensation, *IEEE Transactions on Circuits and Systems for Video Technology*, 7, 560, 1997

[36] Chu, W.-J. and Leou, J.-J., Detection and concealment of transmission errors in H.261 images, *IEEE Transactions on Circuits and Systems for Video Technology*, 8, 74, 1998

[37] Sun, H., Zdepski, J.W., Kwok, W., and Raychaudhuri, D., Error concealment algorithms for robust decoding of MPEG compressed video, *Signal Processing: Image Communication*, 10, 249, 1997

[38] Tsekeridou, S. and Pitas, I., MPEG-2 error concealment based on block-matching principles, *IEEE Transactions on Circuits and Systems for Video Technology*, 10, 646, 2000

[39] Perona, P. and Malik, J., Scale-space and edge detection using anisotropic diffusion, *IEEE Transactions on Pattern Analysis and Machine Intelligence*, 12, 629, 1990

[40] Lee, Y.-C., Altunbasak, Y., and Mersereau, R.M., Multiframe error concealment for MPEG-coded video delivery over error-prone networks, *IEEE Transactions on Image Processing*, 11, 1314, 2002

[41] Lee, S.-H., Choi, D.-H., and Hwang, C.-S., Error concealment using affine transform for H.263 coded video transmissions, *Electronics Letters*, 37, 218, 2001

[42] Atzori, L., De Natale, F.G.B., and Perra, C., A spatio-temporal concealment technique using boundary matching algorithm and mesh-based warping (BMA-MBW), *IEEE Transactions on Multimedia*, 3, 326, 2001

[43] Shirani, S., Kossentini, F., and Ward, R., A concealment method for video communications in an error-prone environment, *IEEE Journal on Selected Areas in Communications*, 18, 1122, 2000

[44] Zhang, Y. and Ma, K.-K., Error concealment for video transmission with dual multiscale Markov random field modeling, *IEEE Transactions on Image Processing*, 12, 236, 2003

[45] Ghanbari, M. and Seferidis, V., Cell-loss concealment in ATM video codecs, *IEEE Transactions on Circuits and Systems for Video Technology*, 3, 238, 1993

[46] Zhang, R., Regunathan, S.L., and Rose, K., Switched error concealment and robust coding decisions in scalable video coding, *Proceedings of the IEEE International Conference on Image Processing*, 3, 2000, 380

[47] Zhu, Q.-F., Wang, Y., and Shaw, L., Coding and cell-loss recovery in DCT-based packet video, *IEEE Transactions on Circuits and Systems for Video Technology*, 3, 248, 1993

[48] Yu, G., Marcellin, M.W., and Liu, M.M.K., Recovery of video in the presence of packet loss using interleaving and spatial redundancy, *Proceedings of the IEEE International Conference on Image Processing*, 2, 1996, 105

[49] Yu, G.-S., Liu, M.M.-K., and Marcellin, M.W., POCS-based error concealment for packet video using multiframe overlap information, *IEEE Transactions on Circuits and Systems for Video Technology*, 8, 422, 1998

[50]   Bystrom, M., Parthasarathy, V., and Modestino, J.W., Hybrid error conceal-
       ment schemes for broadcast video transmission over ATM networks, *IEEE
       Transactions on Circuits and Systems for Video Technology*, 9, 868, 1999
[51]   Robie, D.L. and Mersereau, R.M., Video error correction using steganography,
       Proceedings of the IEEE International Conference on Image Processing, 1, 2001, 930
[52]   Carli, M., Bailey, D., Farias, M., and Mitra, S.K., Error control and concealment
       for video transmission using data hiding, *Proceedings of the Fifth Symposium
       on Wireless Personal Multimedia Communications*, 2, 2002, 812
[53]   Girod, B. and Farber, N., Feedback-based error control for mobile video trans-
       mission, *Proceedings of the IEEE*, 87, 1707, 1999
[54]   Wada, M., Selective recovery of video packet loss using error concealment,
       IEEE Journal on Selected Areas in Communications, 7, 807, 1989
[55]   Wang, J.-T. and Chang, P.-C., Error-propagation prevention technique for
       real-time video transmission over ATM networks, *IEEE Transactions on Circuits
       and Systems for Video Technology*, 9, 513, 1999
[56]   Cen, S. and Cosman, P.C., Decision trees for error concealment in video
       decoding, *IEEE Transactions on Multimedia*, 5, 1, 2003

chapter 7

Image sequence restoration: A wider perspective

Anil Kokaram[1]

Contents

[1] This work was funded by several U.K. and EU projects during 1993–2004.

7.1   Introduction

Within the last 5 years there has been an explosion in the exploitation and availability of digital visual media. Digital television has been widely available in Europe for the last 3 years, and Internet usage continues to grow, as does the availability of MPEG and AVI audio/video clips through the increasing use of streaming media. DVD (digital video/versatile disk) usage is growing faster than CD audio usage did when it was first introduced. There is now growing interest in digital cinema, implying that the whole chain from "film" shooting to distribution and projection will be digital.

With all these available digital video channels, it is amusing to note that the main concern for broadcasters is the relative unavailability of content. Holders of large video, film, and photograph archives — for instance, BBC (U.K.), Institut National de L'Audiovisuel (INA France), and Radio Televisão Portuguesa (RTP Portugal) — find that archive material is in increasing demand. However, the material is typically degraded due to physical problems in repeated projection or playback or simply the chemical decomposition of the original material. Typical problems with much of the archived film material are increased level of noise and dirt and sparkle due to the deposition of dust or the abrasion of the material. Of course, there are many more problems specific to the media, e.g., two-inch tape scratches affecting two-inch videotape and vinegar syndrome and moire affecting film and the film scanning process.

In order to preserve and exploit this material, these defects must be removed so that the picture quality can be restored. Because of the large amount of data, manual retouching is impractical. Therefore, automated techniques have become important. Furthermore, it has been recognized that the reduction of noise, particularly before MPEG compression, allows a more efficient usage of the available digital bandwidth [1, 2, 3].

Hence, the area of automated restoration of image sequences has moved from a principally signal-processing research topic to one of more widespread significance. As an example of increased industrial significance, at the International Broadcast Convention[2] held in Amsterdam in September 2002, there were no fewer than nine companies presenting solutions for automated digital restoration: Phillips, Thomson, Diamant [4], DaVinci, Silicon Graphics, Apple, Discreet, Snell & Wilcox [5], and MTI. Projects such as AURORA (Automated Restoration of Original Film and Video Archives) and BRAVA (Broadcast Restoration of Archives by Video Analysis) funded by the EU are further examples of this increased relevance.

What is more interesting is that with the advent of mobile devices and wireless communications, video transmission over these links is desirable. Such links are so error prone that the problem of reconstructing digitally corrupted data has increased in importance. Furthermore, with the advent of cheap CMOS imagers, pictures taken using mobile devices are sometimes not of good quality, and restoration of those relatively cheaply created images is becoming increasingly important.

It is traditional in books on image processing to find information on restoration that considers the problem of deblurring or denoising images. In the present chapter, the word *restoration* is used in its widest sense to refer to a large range of degradations affecting visual media. Enough information is given here to allow the reader to pursue in more detail the latest advances through the exhaustive references. Most of this chapter deals with a central issue in picture reconstruction: missing data. Before that can be discussed, it is sensible to consider the wide range of defects that exist.

7.1.1 A brief taxonomy of defects

There has been some effort by the BRAVA consortium (http://brava.ina.fr) during 2000–2002 to attempt to standardize or, rather, educate the community about the names and types of defects that can occur in archived video and film. An exhaustive description is given at http://brava.ina.fr/brava_public_impairments_list.en.html. This list was compiled by the project leader, Jean-Hugues Chenot, with input from the various archives in the project, e.g., BBC, INA, and RTP. This kind of taxonomy is notably missing from the recent literature. Figures 7.1–7.10 present part of this taxonomy of defects using the names used in the industry. The methods considered in this chapter address the problems illustrated by Figures 7.1, 7.2, 7.4, and 7.6. Massive loss of data as in Figures 7.3 and 7.7 are better treated through temporal frame interpolation, and the reader may see [6, 7] for treatment of this issue. A possible solution for kinescope moire illustrated in Figure 7.5 can be found in [8, 9]. Two-inch scratches (caused by scratching of old two-inch videotape) are an example of a specialized missing data problem, and a treatment can be found in [10, 11].

[2] One of the two major television technology conferences in the world.

Figure 7.1 Dirt and sparkle occur when material adheres to the film due to electrostatic effects (for instance) and when the film is abraded as it passes through the transport mechanism. This result is also referred to as a blotch in the literature. The visual effect is that of bright and dark flashes at localized instances in the frame. The image indicates where a piece of dirt is visible.

Figure 7.2 Film grain noise is a common effect. It manifests slightly differently depending on the film stock. The image shows clearly the textured visible effect of noise in the sky at the top left. Blotches and noise typically occur together and are the main forms of degradation found on archived film and video. A piece of dirt is indicated on the image.

Figure 7.3 Betacam dropout manifests due to errors on Betacam tape. It is a missing data effect in which several field lines are repeated for a portion of the frame. The repeating field lines are the machine's mechanism for interpolating the missing data.

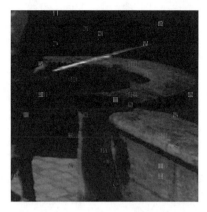

Figure 7.4 Digital dropout occurs because of errors on digital videotape. This example is dropout from D1 tape.

Figure 7.5 Kinescope moire is caused by aliasing during Telecine conversion and manifests as rings of degradation that move slightly from frame to frame.

Figure 7.6 Film tear is simply the physical tearing of a film frame, sometimes due to a dirty splice nearby.

Figure 7.7 Vinegar syndrome often results in a catastrophic breakdown of the film emulsion. This example shows long strands of missing data over the frame.

Figure 7.8 Echoes and overshoots manifest as image shadows slightly displaced from each object in the frame. When the effect is severe, it is called echo. When it is limited to just the edges, as in this case, it is called overshoot.

Figure 7.9 Color fading implies that the picture color is not saturated enough, giving the image a washed-out look.

Two major defects are missing from the visual description: shake and flicker. These are best viewed on video clips. Shake refers to unwanted global motion of the picture caused either by camera movement or problems during scanning. Algorithms for removing shake abound [12, 13, 14, 15, 16, 17]. This is principally because shake is related to the global motion estimation problem that is also important for video-compression issues [18, 19, 20, 21, 22, 23].

Figure 7.10 Line scratches manifest in much archived footage. They also occur due to accidents in film developing. The color of the scratch depends on which side of the film layer it occurs. Often, not all the image information inside the scratch is lost. Line scratches are a challenge to remove because they persist in the same location from frame to frame.

Flicker manifests as a fluctuation in picture brightness from frame to frame. In stills the effect is very difficult to observe, but as a sequence the effect is often very disturbing. Two different types of degradations result in a perceivable flicker artifact. The first realistic de-flicker algorithm was developed by P. M. B. Van Roosmalen et al. [24, 25] and a real-time hardware version was developed by Snell & Wilcox in the late 1990s. Both changing film exposure (in old silent movies, for instance) and varying lighting conditions result in luminance fluctuations. However, misalignment in the two optical paths in a Telecine machine also yields the same visible artifact, called twin lens flicker. In this case, the two fields of each interlaced TV frame are incorrectly aligned with respect to the original film frame, and the observed fluctuations are due more to the shake between the fields than any real luminance changes. Vlachos and Thomas considered this problem in [26] and a real-time implementation was also developed by Snell & Wilcox in the late 1990s.

Video clips showing serious degradation by shake, flicker, lines, grain, and blotches can be seen at www.mee.tcd.ie/~sigmedia/ifc/ifc.html. The book by Read and Meyer [27] gives excellent coverage of the physical nature of archive material and the practices in the film preservation industry.

7.1.2 Missing data treatment in general

This chapter concentrates on a central issue in automated archive restoration: missing data. It manifests as dirt and sparkle (Figure 7.1), dropout (Figures 7.3 and 7.4), film tear (Figure 7.6), dirty splices, and vinegar syndrome (Figure 7.7). Examples of these defects have been illustrated in the previous section. Line scratches can sometimes represent missing data, but quite often data are still in the defect region. Techniques for removing line scratches can be found in [6, 28, 29, 30, 31, 32] and this issue is not considered further here. Missing data deserves special mention as a defect because it occurs also in digital transmission, yielding dropped frames and missing blocks or image bands at the very least. Assuming that the missing data do not occur in the same location in consecutive frames, it seems sensible that repair of the damaged region can be achieved by copying the relevant information from previous or next frames. This relies on the heavy temporal redundancy present in the image sequence. Because this redundancy is prevalent only along motion trajectories, motion estimation has become a key component of missing data treatment systems.

Historically, the approach has been to develop a method to detect the defect [33], and then correct it by spatio-temporal interpolation [34, 35, 36, 37]. The traditional concept in detection is to assume that any set of pixels that cannot be located in next and previous frames must represent an impulsive defect and should be removed. This requires a matching criterion and could be dealt with via a model-based approach [6, 31] or several clever heuristics [6, 31, 38, 39]. Recently, the interaction between the motion estimation and missing data-detection/correction stages has been receiving more attention [40, 41]. A key difference between missing data as it appears in real footage and speckle degradation is that blotches are almost never limited to a single isolated pixel.

It is interesting to note that many of these ideas could be applied to digital images and video corrupted in transmission. However, the video-compression community is not necessarily aware of this parallel work and also tends to rely more on error detection through analysis of the received bitstream syntax. This is not often successful, but it is certainly true that there is even more useful information for restoration and structure in a transmitted compressed bitstream than in a raw image sequence.

The review of missing data treatment therefore begins with a consideration of heuristics for treating this problem. It then moves on to review and extend the model-based approaches. Finally, consideration is given to a major practical shortcoming in the work to date, that of coping with pathological motion.

Much of the work in this area thus far uses a translational motion model for the image sequence as follows

$$I_n(\mathbf{x}) = I_{n-1}(\mathbf{x} + \mathbf{d}_{n,n-1}(\mathbf{x})) + e(\mathbf{x}) \tag{7.1}$$

where the luminance at pixel site $\mathbf{x} = [i,j]$ in frame n is $I_n(\mathbf{x})$, and the two-component motion vector mapping site $\mathbf{x}$ in frame n into frame $n-1$ is $\mathbf{d}_{n,n-1}(\mathbf{x})$. The model error follows a Gaussian distribution $e(\cdot) \sim \mathcal{N}(0,\sigma_e^2)$ and is consistent with luminance-conserving, translational motion in the sequence. Figure 7.11 illustrates these concepts.

7.2   Heuristics for detection

Perhaps the earliest effort to "electronically" detect dirt and sparkle was undertaken by Richard Storey at the BBC in 1983 [38, 42]. The idea was to flag a pixel as missing if the forward and backward pixel difference was high. This effort was, of course, beset with problems in parts of the image where motion occurred. The natural extension of this idea was presented by

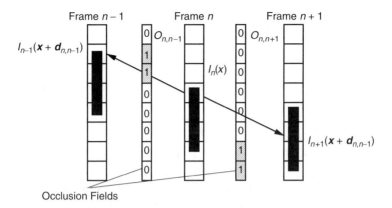

Figure 7.11 Using three 1-D image frames to illustrate the notation used in the chapter to indicate a pixel value $I_n(\mathbf{x})$ at position $\mathbf{x}$ in frame n and its motion-compensated correspondences in frames $n-1$ and $n-2$. The motion between the frames is $\mathbf{d}_{n,n+1}(\mathbf{x})$ and $\mathbf{d}_{n,n-1}(\mathbf{x})$ in the forward and backward directions, respectively. The motion vectors contain two components for horizontal and vertical displacement. The object traverses four pixels and is shown as a block bar. Occlusion fields between the observed frames indicate when pixels are occluded between frames. Thus a pixel in an occlusion field $O_{n,n-1}$ or $O_{n,n+1}$ is set to 1 if the corresponding pixel in the image n is not observed in the previous frame $(n-1)$ or the next frame $(n+1)$.

Kokaram around 1993 [6, 33], which allowed for motion compensated differences. That type of detector was called a Spike Detection Index (SDI), and the most useful are defined as follows.

7.2.1 SDIx

The forward and backward motion-compensated pixel differences F_f and E_b of the observed corrupted image sequence $G_n(\mathbf{x})$ are defined as

$$E_b = G_n(\mathbf{x}) - I_{n-1}\big(\mathbf{x} + \mathbf{d}_{n,n-1}(\mathbf{x})\big)$$

$$E_f = G_n(\mathbf{x}) - I_{n+1}\big(\mathbf{x} + \mathbf{d}_{n,n-1}(\mathbf{x})\big)$$

(7.2)

Note that the previous and next frames are assumed to be uncorrupted at the required motion-compensated sites; hence, $G_{n-1} = I_{n-1}$ etc. Two detectors can then be proposed [6] as

$$b_{\mathrm{SDIa}}(\mathbf{x}) = \begin{cases} 1 & \text{for } \big(|E_b| > E_t\big) \text{ AND } \big(|E_f| > E_t\big) \\ 0 & \text{otherwise} \end{cases}.$$

(7.3)

$$b_{\text{SDIp}}(\mathbf{x}) = \begin{cases} 1 & \text{for } \left(|E_b| > E_t\right) \text{ AND } \left(|E_f| > E_t\right) \\ & \text{AND } \text{sign}(E_f) = \text{sign}(E_b) \\ 0 & \text{otherwise} \end{cases} \qquad (7.4)$$

Here, b(·) is a detection field variable set to 1 at sites that are corrupted by missing data. Figure 7.12 shows the relationship of the pixels used to each other. The Spike Detection Index a (SDIa) is based on thresholding E_f and E_b only. The Spike Detection Index p (SDIp) additionally applies the constraint that if corruption does not occur in identical locations in consecutive frames and the model in Equation (7.1) holds, $I_{n-1} \approx I_{n+1}$, then one should expect that the sign of the difference signals should be the same. It is now accepted that SDIp is the better detector in almost all situations because of this additional constraint.

7.2.2 ROD

In 1996, Nadenau and Mitra [39] presented another scheme that used a spatio-temporal window for inference: the rank order detector (ROD). It is generally more robust to motion-estimation errors than any of the SDI detectors, although it requires the setting of three thresholds. It uses some spatial information in making its decision. The essence of the detector is the premise that

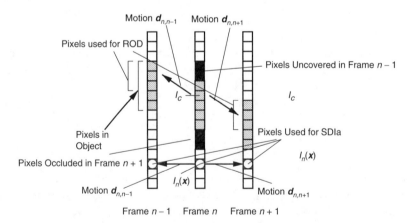

Figure 7.12 Using three 1-D image frames to illustrate the pixels used in the SDIx and ROD detectors. Both use motion-compensated pixels, but the ROD uses a spatio-temporal neighborhood of six pixels whereas the SDI detectors rely on just a three point neighborhood. The SDI detectors are therefore more sensitive to errors in motion estimation.

blotched pixels are outliers in the local distribution of intensity. The pixels used in the ROD are shown in Figure 7.12, and the algorithm is as follows:

1. Sort the 6 motion-compensated pixels into a list $[r_1, r_2, r_3, \cdots, r_6]$ where r_1 is minimum. The median of these pixels is then calculated as $M = (r_3 + r_4)/2$.
2. Three motion-compensated difference values are calculated as follows:

$$\text{If } I_c > M \qquad\qquad\qquad\qquad \text{If } I_c \leq M$$

$$\begin{aligned} e_1 &= I_c - r_6 & e_1 &= r_1 - I_c \\ e_2 &= I_c - r_5 & e_2 &= r_2 - I_c \\ e_3 &= I_c - r_4 & e_3 &= r_3 - I_c \end{aligned}$$

3. Three thresholds are selected: t_1, t_2, and t_3. If any of the differences exceeds these thresholds, then a blotch is flagged as follows:

$$b_{\text{ROD}}(\mathbf{x}) = \begin{cases} 1 & \text{if } (e_1 > t_1) \text{ OR } (e_2 > t_2) \text{ OR } (e_3 > t_3) \\ 0 & \text{otherwise} \end{cases}$$

where $t_3 \geq t_2 \geq t_1$. The choice of t_1 is the most important. The detector works by measuring the "outlierness" of the current pixel when compared to a set of others chosen from other frames. The choice of the shape of the region from which the other pixels were chosen is arbitrary.

7.2.3   Image histograms

Many authors have used the image histogram to detect abrupt changes in image sequences. This idea can be extended to detect very large areas of missing data in degraded archive frames. It is totally unsuitable for detection of all but the largest defects since otherwise the changes in consecutive histograms are not noticeable. In [43, 44] Kokaram et al. present a mechanism for isolating the image quadrant that contains the defect. The attractive aspect of this proposal is that no motion estimation is necessary; hence, the computational cost is extremely low. However, it is a very coarse detection process and is not examined further here. Such a coarse detection process can be used by restoration operators to browse a video sequence for detection of the most severely damaged frames.

7.2.4   Morphological/median filter approaches

In the 1-D case, Paisan and Crise [45] were the first to spot that one could use a median-filtered signal as a *rough estimate* of the original signal before corruption by impulsive noise. The difference between the observed, degraded signal and this rough estimate would be high at sites corrupted by impulsive defects. This is because the rank order filter removes all outliers but leaves lower scale trends untouched. This idea can be extended to treat small missing data artifacts in archived video and film, known as *dust*. These are generally just a few pixels in area (3× 3 pixels); hence, only a small median or morphological window need be used. Using a larger window to detect larger artifacts causes problems since more true image detail would then be removed, causing an increased number of false alarms. Researchers have been implementing these types of techniques for film restoration since the mid-1990s [31, 46–54]. Joyeux et al. [31] point out that these techniques are particularly attractive because of their low computational cost. They perform well when the artifact is small and surrounded by a relatively low-activity homogenous region. The high resolution of film scans is therefore suitable for these tools.

7.3   Missing data reconstruction

Having detected an area of missing data, it is necessary to synthesize material to fill the gap. In 1983, Storey [38, 42] used a three tap median operation to interpolate the missing pixel. This was implemented without motion information. In 1993, Kokaram extended the idea to motion-compensated filtering but, recognizing that motion-estimation errors resulted in very poor performance, introduced a three-dimensional median filtering operation on a 3× 3 × 3 motion-compensated pixel volume around each missing pixel [34]. A model-based scheme that used 3-D AR models to synthesize texture in the gap was also presented in [34]. Model-based pixel cut-and-paste operations from previous or next frames then followed, which allowed for occlusion and uncovering. Deterministic frameworks were proposed circa 1996 [55, 56], and a Bayesian cut-and-paste method was proposed by Van Roosmalen et al. [36] in 1999. The essence of all these ideas was to ensure that interpolated pixel data was smooth both in time and space.

Since 2000 there has been an increasing interest in the concept of spatial filling in of gaps in images. The term *inpainting* was used to describe this idea by Ballester and Bertalmio et al. [57, 58]. In 2002 similar ideas emerged based on 2-D AR interpolation [59, 60]. The remarkable results of Efros and Freeman [61, 62] have also received considerable attention. One could consider that these techniques can be used to reconstruct small missing data regions in a degraded image. Bornard [63, 64], in his thesis of 2002, has considered using these techniques as an alternative to volumetric image data reconstruction because of the problems associated with motion estimation in difficult situations. Remarkable results have been achieved on relatively small corruption without the need for motion interpolation.

Spatial morphological operators as well as deterministic spatial interpolation can be used to good effect to reconstruct missing data for small blotches (called dust). However, these methods tend to work best when the region to be reconstructed is relatively homogenous [31]. For high-resolution film applications in particular, these methods are quite appropriate for handling many of the smaller defects.

In the interest of brevity, an exhaustive comparison of spatial versus spatio-temporal interpolation is not undertaken here. It is important to note, however, that for large missing regions, e.g., half a frame, spatial interpolation is unable to cope, while for small regions, especially in homogenous areas, spatial interpolation is ideal.

7.4 The Bayesian approach

One problem with the aforementioned approaches to missing data treatment is that they address single issues separately and do not acknowledge the interactions among their operations. For instance, all of the detectors noted cause a high rate of false alarms in the presence of poor motion estimation. However, the presence of missing data can cause poor motion-estimation performance. In addition, reconstruction using any of the volumetric techniques, especially for large pieces of missing data, is poor when motion estimation is poor. Poor detection generally leads to poor reconstruction since the detector is probably flagging areas that cannot be properly reconstructed. To give good performance, therefore, the motion estimate must be robust to the presence of missing data, and the detector should work in tandem with the interpolator in some way to evaluate whether more damage is being introduced than removed.

Another problem is that none of the approaches incorporates all that is known about the nature of the defects in question. For instance, missing regions are generally smooth areas that are spatially coherent in some sense. They also usually do not occur in the same place in two consecutive frames. There are indeed many postprocessing schemes that can achieve this information insertion via clever filtering operations or morphological dilation/ erosion pairs [6, 35], but it is useful to consider how this information could be incorporated into the problem definition itself.

Model-based schemes could provide one solution to this issue, and schemes for missing data detection and interpolation were reviewed and compared with deterministic approaches in [6, 33, and 34]. In 1994, Morris [65] was the first to propose using the Bayesian approach to resolve this data fusion aspect of combining different sorts of prior information about blotches. The Bayesian approach has since been evolving [6, 36, 41, 59, 66, 67] into a unifying framework that treats motion, missing data, and noise jointly. Reference [6] contains an exhaustive discussion of a joint detection/ interpolation scheme for blotches.

The main problems with the previous approaches presented in [6 and 41] are computational complexity and the failure in cases of motion discontinuity. This can be solved by unifying the problems of motion estimation and picture reconstruction with occlusion estimation. The next sections present this idea within a Bayesian framework.

7.4.1　Quantifying the problem

There are two effects to be modeled as far as missing data is concerned, the missing data itself (data that replaces the existing data) and random additive noise. The random additive noise $\mu(\mathbf{x})$ can be modeled as a Gaussian i.i.d. process with $\mu(\cdot) \sim \mathcal{N}(0, \sigma_\mu^2)$. Modeling the missing data involves two ideas: the first is modeling the *location* of the corruption; the second is modeling the actual *nature* of the corruption. To model location, a binary field $b(\mathbf{x})$ is introduced that is 1 at a site of missing data and zero otherwise. To model the corruption (the look and feel of blotches) a texture field $c(\mathbf{x})$ is introduced that *replaces* the image data wherever $b(\mathbf{x}) = 1$. Therefore, the observed, dirty frames G_n can be created by mixing the clean image I_n with noise and a blotch corruption layer $b(\mathbf{x})c(\mathbf{x})$.

The degradation model can then be written as

$$G_n(\mathbf{x}) = \left(1 - b(\mathbf{x})\right) + I_n(\mathbf{x}) + b(\mathbf{x})c(\mathbf{x}) + \mu(\mathbf{x}) \tag{7.5}$$

where $\mu(\cdot) \sim \mathcal{N}\left(0, \sigma_\mu^2\right)$ is the additive noise and $c(\mathbf{x})$ is a field of random variables that cause the corruption at sites where $b(\mathbf{x}) = 1$. This formulation models the degradation of the clean images and, therefore, μ is not the same as e in Equation (7.1). The noise in Equation (7.1) attempts to quantify the uncertainty in the image sequence model that relates clean frames to each other. Unfortunately, it is extremely difficult to keep these two effects separate in any solution to this problem.

Three distinct (but interdependent) tasks can now be identified in the restoration problem. The ultimate goal is image estimation, i.e., revealing $I_n(\cdot)$ given the observed missing and noisy data. The missing data detection problem is that of estimating $b(\mathbf{x})$ at each pixel site. The noise reduction problem is that of reducing $\mu(\mathbf{x})$ without affecting image details. The replacement model was employed within a nonprobabilistic framework by Kokaram and Rayner [68] for image sequences, and it was implicitly employed in a two-stage Bayesian framework for missing data detection and interpolation by Morris et al. [33, 69].

7.4.2　The image sequence model: pixel states and occlusion

To incorporate occlusion into the image sequence model in Equation (7.1), a hidden field of binary variables $O_b(\mathbf{x})$ and $O_f(\mathbf{x})$ is introduced between frames $n, n - 1$ and $n, n + 1$, respectively. When $O_b(\mathbf{x}) = 1$, this implies that the data at site $\mathbf{x}$ in frame n does not exist in the frame $n - 1$. This represents occlusion in the backward temporal direction. A similar situation exists in the forward direction with $O_f(\mathbf{x})$. This idea is illustrated in Figure 7.11 as part of the image sequence model statement.

Degradation information can be included in the pixel site information by defining a pixel as occupying one of six states. Each of these states $s(\mathbf{x}) \in [S1 \dots S6]$ is defined as a combination of three binary variables $[b(\mathbf{x}), O_b(\mathbf{x}), O_f(\mathbf{x})]$ as follows.

001 The pixel is not missing and there is occlusion in the forward direction only.

010 The pixel is not missing and there is occlusion in the backward direction only.

000 The pixel is not missing and there is no occlusion backward or forward.

100 The pixel is corrupted by a blotch and there is no occlusion backward or forward.

101 The pixel is corrupted by a blotch and there is occlusion in the forward direction only.

110 The pixel is corrupted by a blotch and there is occlusion in the backward direction only.

Note that in this framework the [1 1 1] state is not allowed since it would imply that the data is missing and yet there is no temporal information for reconstruction. This is an interesting practical omission, and some comments are made at the end of the chapter.

7.4.3 A Bayesian framework

From the degradation model of (7.5), the principal unknown quantities in frame n are $I_n(\mathbf{x})$, $s(\mathbf{x})$, $c(\mathbf{x})$, the motion $\mathbf{d}_{n,n-1}$ and the model error $\sigma_e^2(\mathbf{x})$. These variables are lumped together into a single vector $\boldsymbol{\theta}(\mathbf{x})$ at each pixel site $\mathbf{x}$. The Bayesian approach presented here infers these unknowns conditional upon the corrupted data intensities from the current and surrounding frames $G_{n-1}(\mathbf{x})$, $G_n(\mathbf{x})$, and $G_{n-1}(\mathbf{x})$. For the purposes of missing data treatment, it is assumed that corruption does not occur at the same location in consecutive frames; thus, in effect, $G_{n-1}=I_{n-1}$, $G_{n+1}=I_{n+1}$.

Proceeding in a Bayesian fashion, the conditional may be written in terms of a product of a likelihood and a prior as

$$p\big(\boldsymbol{\theta}\big|I_{n-1},G_n,I_{n+1}\big) \propto p\big(G_n\big|\boldsymbol{\theta},I_{n-1},I_{n+1}\big)p\big(\boldsymbol{\theta}\big|I_{n-1},I_{n+1}\big).$$

This posterior may be expanded at the single pixel scale, exploiting conditional independence in the model, to yield

$$p(\boldsymbol{\theta}(\mathbf{x})\,|\,G_n(\mathbf{x}),I_{n-1},I_{n+1},\boldsymbol{\theta}(-\mathbf{x}))$$

$$\propto p(G_n(\mathbf{x})\,|\,\boldsymbol{\theta}(\mathbf{x}),I_{n-1},I_{n+1})p(\boldsymbol{\theta}(\mathbf{x})\,|\,I_{n-1},I_{n+1},\boldsymbol{\theta}(-\mathbf{x}))$$

$$= p(G_n(\mathbf{x})\,|\,I_n(\mathbf{x}),c(\mathbf{x}),b(\mathbf{x})) \times p(I_n(\mathbf{x})\,|\,s_e(\mathbf{x})^2,\mathbf{d}(\mathbf{x}),O_b(\mathbf{x},O_f(\mathbf{x}),I_{n-1},I_{n+1}))$$

$$\times p(b(\mathbf{x})\,|\,B)p(c(\mathbf{x})\,|\,C)p(\mathbf{d}(\mathbf{x})\,|\,D)p(s_e(\mathbf{x})^2)\,p(O_b(\mathbf{x})\,|\,O^b)p(O_f(\mathbf{x})\,|\,O^f)$$

$$(7.6)$$

where $\theta(-\mathbf{x})$ denotes the collection of θ values in frame n with $\theta(\mathbf{x})$ omitted and B, C, D, O^b, O^f, and I denote local dependence neighborhoods around $\mathbf{x}$ (in frame n) for variables b, c, $\mathbf{d}$, O_b, O_f, and I_n, respectively. See the sections on prior distributions for details of these neighborhoods. To proceed, precise functional forms for the likelihoods must be assigned.

7.4.4 The corruption likelihood

Considering Equation (7.5), $p(G_n | \cdot)$ has different forms depending on the state of $b(\mathbf{x})$, as (dropping the argument $\mathbf{x}$ for clarity)

$$p(G_n | I_n, c, b) \propto \begin{cases} \exp-\left(\dfrac{(G_n - I_n)^2}{2s_m^{\,2}}\right) & b = 0 \\[4mm] \exp-\left(\dfrac{(G_n - c)^2}{2s_m^{\,2}}\right) & b = 1 \end{cases} \tag{7.7}$$

7.4.5 The original (clean) data likelihood

This expression is derived directly from the image sequence model and also involves interaction with the occlusion variables:

$$p\left(I_n(\mathbf{x}) | I_{n-1}, I_{n+1}\right) \propto \exp-\left(\frac{\left(1 - O_b(\mathbf{x})\right)I_n(\mathbf{x}) - I_{n-1}\left(\mathbf{x} + \mathbf{d}_{n,n-1}\right)\right)^2}{2\sigma_e^2}\right)$$

$$\times \exp-\left(\frac{\left(1 - O_f(\mathbf{x})\right)\left(I_n(\mathbf{x}) - I_{n+1}\left(\mathbf{x} + \mathbf{d}_{n,n+1}\right)\right)^2}{2\sigma_e^2}\right) \times \exp\left(-\alpha O_b(\mathbf{x})\right)\exp\left(-\alpha O_f(\mathbf{x})\right).$$

$$\tag{7.8}$$

The likelihood is therefore proportional to the displaced frame difference (DFD), defined as $I_n(\mathbf{x}) - I_{n-1}(\mathbf{x} + \mathbf{d}_{n,n-1})$ between the image frames in both temporal directions. This expression encourages smoothness in intensities along a motion trajectory. However, when occlusion occurs (e.g., $O_b(\mathbf{x}) = 1$), the DFD is turned off because there is no valid data in the preceding (or next) frame.

This expression alone would yield the degenerate solution of $O_b = O_f = 1.0$ everywhere in the image because that would maximize the impact of this likelihood on the posterior. Therefore, it becomes necessary to introduce a penalty for setting occlusion to the ON state. This is represented by $\alpha O_b(\mathbf{x})$ where $\alpha = 2.76^2/2.0$. This value is chosen to represent roughly the 90% confidence limit for a Gaussian distribution. See [6] for further justification.

It is now necessary to specify priors for the variable θ.

7.4.6 Priors

The remaining distributions encode the prior belief about the values of the various unknowns. The variance σ_e^2 is assigned a noninformative prior $p(\sigma_e^2) \propto 1/\sigma_e^2$, following [70].

7.4.6.1 Motion

The prior adopted for motion smoothness is a Gibbs Energy prior, for instance as introduced by Konrad and Dubois [71]. To reduce the complexity of the final solution, the motion field is block based with one motion vector employed for each specified block in the image. The prior for $\mathbf{d}_{n,n-1}(\mathbf{x})$, the motion vector mapping the pixel at $\mathbf{x}$ in frame n into frame $n-1$, is

$$p_d(\mathbf{d}_{n,n-1}(\mathbf{x}) \,|\, \mathbf{d}_{n,n-1}(-\mathbf{x}), \mathbf{V}(\mathbf{x})) \propto \exp-\left(\sum_{\mathbf{x}_v \in \mathbf{V}(\mathbf{x})} \lambda(\mathbf{x}_v)[\mathbf{d}_{n,n-1}(\mathbf{x}) - \mathbf{d}(\mathbf{x}_v)]^2 \right) \quad (7.9)$$

where $\mathbf{x}_v$ is one site in the eight-connected block neighborhood represented by $\mathbf{V}(\mathbf{x})$, and $\lambda(\mathbf{x}_v)$ is the weight associated with each clique. The same prior is used for $\mathbf{d}_{n,n+1}(\mathbf{x})$. In order to discourage "smoothness" over too large a distance, $\lambda(\mathbf{x}_v)$ is defined with $\lambda(\mathbf{x}_v) = \Lambda / |\mathbf{X}(\mathbf{x}_v) - \mathbf{x}_B|$ where $\mathbf{X}(\mathbf{x}_v)$ is the location of the block (in block units) providing the neighborhood vector $\mathbf{d}(\mathbf{x}_v)$, and $\mathbf{x}_B$ is the central block location. $\Lambda = 2.0$ in the experiments presented later.

7.4.6.2 The priors for corruption, detection, and occlusion

Since blotches tend to be "convex" clumps of degradation, the prior for $b(\mathbf{x})$ should encourage contiguous areas of $b = 1$ to form. In practice, blotches tend to have fairly constant intensity (see Figures 7.1 and 7.13). If a texture exists, it is certainly smoother than that in the original image. Thus, the prior for $c(\mathbf{x})$ should encourage smoothness of intensity, in much the same way that the prior for $b(\cdot)$ encourages smoothness in its binary configuration. Therefore, it is reasonable to place a similar energy prior on both the binary field $b(\mathbf{x})$ and the blotch value field $c(\mathbf{x})$. A GMRF (Gaussian Markov Random Field) prior is used for $c(\mathbf{x})$. The prior for $b(\mathbf{x})$ is similar but operates on a binary variable only; thus, it is related to the Ising model.

The priors are therefore defined as

$$p_c(c(\mathbf{x}) \,|\, C) \propto \exp\left(-\sum_{k=1}^{8} \lambda_k^c (1 - u(\mathbf{x}, \mathbf{x} + \mathbf{v}_k))(c(\mathbf{x}) - c(\mathbf{x} + \mathbf{v}_k)^2 \right) \quad (7.10)$$

$$p_b(b(\mathbf{x}) \,|\, B) \propto \exp\left(-\sum_{k=1}^{8} \lambda_k^b (1 - u(\mathbf{x}, \mathbf{x} + \mathbf{v}_k)) \,|\, b(\mathbf{x}) - b(\mathbf{x} + \mathbf{v}_k)| \right) \quad (7.11)$$

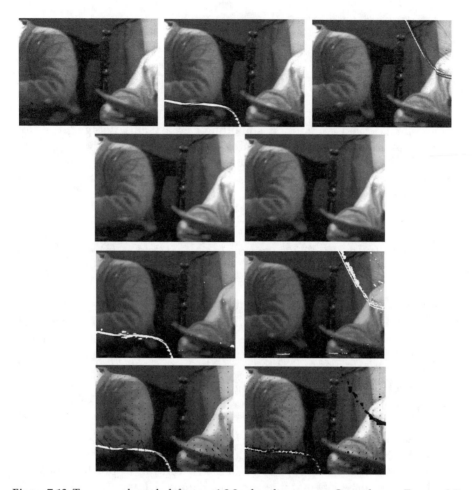

Figure 7.13 Top row: degraded frames 1,2,3 of real sequence; Second row: Frames 2,3 restored with algorithm described in this chapter; Third row: $b(\mathbf{x})$ for frames 2,3 superimposed on darkened original frames; Bottom row: initial motion estimate (left) superimposed on $c(\mathbf{x})$ for frame 2. Motion field after five iterations of algorithm together with $O_b('o')$, $O_f('x')$.

where the eight offset vectors $\mathbf{v}_k$ define the eight-connected pixel neighborhood of $\mathbf{x}$, and C, B represent sets of these values from the respective fields (as previously). The situation is illustrated in Figure 7.14 using a typical eight-connected neighborhood set for v_k. $u(\mathbf{x}, \mathbf{x} + \mathbf{v}_k)$ is set to 1 if there is a significant zero crossing between the location $\mathbf{x}$ in the image and $\mathbf{x} + \mathbf{v}_k$ from which a neighborhood pixel is extracted. Thus, the smoothness constraint is turned off across significant edges in the image. Note that these priors are defined on the pixel resolution image grid, whereas the motion prior discussed previously is defined on a block grid. In the results shown later, $u(\cdot)$ (the edge field) was configured using an edge detector employing difference of Gaussians with the gradient threshold set at 5.0, the variance of the Gaussian filters was 1.0, 1.6, and the filter window sizes were 11×11.

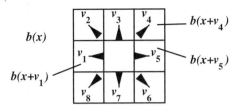

Figure 7.14 The spatial dependence in the prior for blotches and corruption.

For both of these priors, λ_k^c, λ_k^b are assigned values such that $\lambda_k^c = \Lambda^c / |\mathbf{v}_k|, \lambda_k^b = \Lambda^b / |\mathbf{v}_k|$. This makes the hyperparameter weighting circularly symmetric. Typical values are $\Lambda^c = 0.15$, $\lambda^b = 4.0$. These are found from experiment to yield good results over a wide cross-section of degraded sequences.

The occlusion priors $p_o(O_b|\mathbf{O}_b), p_o(O_f|\mathbf{O}_f)$ are identical to the prior for the blotch indicators b, with $\Lambda_o = 2.0$. It encourages organization (clumpiness) in the occlusion surfaces. This is sensible given that the moving objects tend to cause contiguous regions of occlusion and uncovering.

7.4.7 Solving for the unknowns

The solution is generated by manipulating $p(\boldsymbol{\theta}|\,I_{n-1},G_n,I_{n+1})$. For instance, the maximum *a posteriori* estimate is generated by maximizing the distribution with respect to the unknowns. Unfortunately, due to the nonlinear nature of the expression, a closed-form solution to the optimization problem is not available. To yield a practical solution, a number of simplifying manipulations will be made.

It is expedient at this stage to note that, in fact, the variables of this problem can be grouped into a pixel state, $s(\mathbf{x}) = [b(\mathbf{x}), O_b(\mathbf{x}), O_f(\mathbf{x})]$ (defined previously), and its associated "clean" image value $I_n(\mathbf{x})$, at each pixel site, and a motion and error variable

$$\left[\mathbf{d}_{n,n-1},\mathbf{d}_{n,n+1},\sigma_e^2\right]$$

for each block in the image. In other words, the variables can be grouped into a pixelwise group $[s,I_n,c]$ (including the corruption value c) that varies on a pixel grid and a blockwise group

$$\left[\mathbf{d}_{n,n-1},\mathbf{d}_{n,n+1},\sigma_e^2\right]$$

that varies on a coarser grid.

To solve for these unknowns, the iterated conditional modes (ICM) algorithm [72] is used. At each site, the variable set that maximizes the local conditional distribution, given the state of the variables around, is chosen as a suboptimal estimate. Each site is visited with variables being replaced with ICM estimates. By treating each group of variables jointly, an efficient solution results. Thus, first the group $[s, I_n, c]$ is estimated given the current estimate for motion, and then the blockwise motion information is estimated given the current reconstructed image estimate I. This process is then iterated.

7.4.7.1 Factoring

It transpires that $s, I_n(\mathbf{x})$ and $\sigma_e^2, \mathbf{d}$ can be *jointly* manipulated by factoring the posterior. To illustrate, the p.d.f. for $s, I_n(\mathbf{x})$ can be factored by the decomposition

$$p(s, I(\mathbf{x}) \mid I_{n-1}, I_{n+1}, \mathbf{d}, \sigma_e^2) = p(I(\mathbf{x}) \mid s, I_{n-1}, I_{n+1}, \mathbf{d}, \sigma_e^2) p(s \mid I_{n-1}, I_{n+1}, \mathbf{d}, \sigma_e^2). \quad (7.12)$$

The algorithm therefore proceeds by first solving for s with $p(s \mid I_{n-1}, I_{n+1}, \mathbf{d}, \sigma_e^2)$ and then using that estimate in the generation of the clean image data. The various factor terms can be derived by successively *integrating out* $I_n(\mathbf{x})$ and $c(\mathbf{x})$ from the posterior distribution. See [73] for details. Thus, the state s that maximizes $p(s \mid I_{n-1}, I_{n+1}, \mathbf{d}, \sigma_e^2)$ is chosen first. Then that value is used in estimating $I(\mathbf{x})$ from $p(I(\mathbf{x}) \mid s, I_{n-1}, I_{n+1}, \mathbf{d}, \sigma_e^2)$.

Although suboptimal, the approximation is helped by the observation that there is no spatial dependence in the conditional for the clean image data $p(I_n(\mathbf{x}) \mid \cdot)$ and the conditional $p(\sigma_e^2(\mathbf{x}) \mid \cdot)$. Maximizing the local conditional distribution for s can be performed very efficiently by evaluating the posterior for the six possible state options and simply choosing that which maximizes the posterior. The derivation of the state conditionals is tedious. The reader may refer to [73] for details.

7.4.7.2 Manipulating motion

Recall that in this implementation, motion is handled on a block basis. Integration of the posterior yields the following factorization of $p(\sigma_e^2, \mathbf{d} \mid \cdot)$, using the backward motion $\mathbf{d}_{n,n-1}$ and backward frame pair

$$p(\sigma_e^2 \mid \mathbf{d}_{n,n-1}, \mathbf{i}) = IG(N/2, E(\mathbf{i}, \mathbf{d}_{n,n-1})/2)$$

$$p(\mathbf{d}_{n,n-1} \mid \mathbf{i}, D) \propto E(\mathbf{i}, \mathbf{d}_{n,n-1})^{-N/2} p_d(\mathbf{d}_{n,n-1} \mid D) \quad (7.13)$$

where IG denotes the inverted gamma distribution, N is the number of pixels in the image block, D (shorthand for $\mathbf{S}_n(\mathbf{x})$) represents a neighborhood of vectors surrounding the current location, and $E(\mathbf{i}, \mathbf{d}_{n,n-1})$ is the sum of the square of the DFDs in a block (allowing for occlusion) given the vector $\mathbf{d}$. The expressions for the forward frame pair are similar except for the use of $\mathbf{d}_{n,n+1}$ instead. Note that the distinction between forward and backward temporal directions means that the forward and backward DFD variance is allowed to vary. The derivation of these expressions can be found in [6, 73].

Again the local conditional is maximized in parts, first $\mathbf{d}$ from $p(\mathbf{d} \mid \mathbf{i}, D)$, then σ_e^2 from maximizing the IG distribution above. The maximum of the

IG distribution can be calculated analytically and is simply the usual estimate for variance given **d** as

$$\hat{\sigma}_e^2 = \frac{1}{N} \sum_{\mathbf{x} \in B} \left[\text{DFD}(\mathbf{x}) - \overline{\text{DFD}(\mathbf{x}, \mathbf{d})} \right]^2 \qquad (7.14)$$

where $\overline{\text{DFD}(\mathbf{x})}$ is the mean of the DFD in that block of pixels.

The maximization of the motion conditional has no straightforward analytic solution. However, a simple and practical solution is to use fast deterministic motion estimation schemes to yield candidates for motion, for example, block matching or optic flow schemes. The marginal conditional probability of these candidates is then evaluated using Equation (7.13). This idea has not been fully explored in the literature but does yield much practical benefit.

7.4.7.3 The final algorithm

The algorithm begins with the initialization of the fields using deterministic estimates. Motion is initialized using some standard block-based motion estimator. The detection field b is initialized using any one of the deterministic schemes, e.g., SDIp, ROD, or morphological operations [31]. A conservative initial estimate for $c(\mathbf{x})$ is $G_n(\mathbf{x})$. It then proceeds on the basis of two major iterations:

1. Given the current estimates for $\mathbf{d}, \sigma_e^2$, the image is swept on a pixel basis, using a checkerboard scan, generating an estimate for s and $I(\mathbf{x})$ at each site.
2. Given the current estimates for s and $I(\mathbf{x})$, the image is scanned on a block basis. Within each block, the solution for motion is generated by selecting eight candidate motion vectors from surrounding spatial blocks and choosing the one that minimizes the following "energy" (arising from Equation (7.13))

$$\frac{N}{2} \log_e \left[\sum_{\mathbf{x} \in B} (I_n(\mathbf{x}) - I_{n-1}(\mathbf{x} + \mathbf{d})(1 - O_b(\mathbf{x}))^2 \right] + \log_e (p_d(\mathbf{d}|\cdot)). \qquad (7.15)$$

σ_e^2 is then measured as the variance of the prediction error in the block after the motion is selected. Note that as the occlusion variables are taken into account (removing pixel pairs that are occluded), the normalizing term $(N/2)$ changes accordingly. In cases where there is no valid temporal data, $\log_e (p_d(\mathbf{d}|\cdot))$ dominates, and the smoothest motion field will be selected regardless of DFD.

These two processes are iterated until reasonable pictures are built. It is found that typically no more than 10 such iterations are needed for good pictures to be built. The algorithm for the selection of the best state s at each pixel is as follows.

7.4.7.4   Maximizing with respect to S

At each pixel site there are just six possibilities for states. Each of these states defines a value for all the pixelwise variables under consideration. By integrating out c and then $I_n(x)$ from the posterior (see [6]), the maximization process manifests as the minimization of six possible energies (log likelihoods) depending on the values of state. Four of these are shown below. Details are in [73]. Note that g_n, i_n is shorthand for a single pixel observation of the degraded and clean image, respectively, i.e., $G_n(x), i_n$. In the expressions that follow, it is assumed that any data required from frames $n - 1$, $n + 1$ are compensated for motion.

$$\mathcal{E}(S1) = \log_e\left[2\pi\sigma_e^2\sigma_\mu^2\right] + \alpha + \left[\frac{\left(g_n - \hat{i}_b\right)^2}{2\sigma_\mu^2}\right] + \left[\frac{\left(i_{n-1} - \hat{i}_b\right)^2}{2\sigma_e^2}\right]$$

$$+ \log_e\left[p_b(b = 0|B)p_o\left(O_b = 0|\mathbf{O_b}\right)p_o\left(O_f = 1|\mathbf{O_f}\right)\right] + \log_e\left[p_c\left(c = \hat{c}_0|C\right)\right]$$

$$\mathcal{E}(S3) = \frac{1}{2}\log_e\left[8\pi^3\sigma_e^4\sigma_\mu^2\right] + \left[\frac{\left(g_n - \hat{i}\right)^2}{2\sigma_\mu^2}\right] + \left[\frac{\left(i_{n-1} - \hat{i}\right)^2}{2\sigma_e^2}\right] + \left[\frac{\left(i_{n+1} - \hat{i}\right)^2}{2\sigma_e^2}\right]$$

$$+ \log_e\left[p_b(b = 0|B)p_o\left(O_b = 0|\mathbf{O_b}\right)p_o\left(O_f = 0|\mathbf{O_f}\right)\right] + \log_e\left[p_c\left(c = \hat{c}_0|C\right)\right]$$

$$\mathcal{E}(S4) = \frac{1}{2}\log_e\left[2\pi\sigma_\mu^2\right] + \left[\frac{\left(g_n - \hat{c}\right)^2}{2\sigma_\mu^2}\right]$$

$$+ \log_e\left[p_b(b = 1|B)p_o\left(O_b = 0|\mathbf{O_b}\right)p_o\left(O_f = 0|\mathbf{O_f}\right)\right] + \log_e\left[p_c\left(c = \hat{c}_1|C\right)\right]$$

$$\mathcal{E}(S6) = \frac{1}{2}\log_e\left[2\pi\sigma_\mu^2\right] + \alpha + \left[\frac{\left(g_n - \hat{c}\right)^2}{2\sigma_\mu^2}\right]$$

$$+ \log_e\left[p_b(b = 1|B)p_o\left(O_b = 1|\mathbf{O_b}\right)p_o\left(O_f = 0|\mathbf{O_f}\right)\right] + \log_e\left[p_c\left(c = \hat{c}_1|C\right)\right]$$

Here i_{n-1} denotes a motion-compensated pixel in frame $n - 1$. The various constants $\hat{i}_b, \hat{i}_f, \hat{i}, \hat{c}$ are the least squared estimates of the unknowns $I_n(x), c(x)$ given various temporal situations defined by the state variable, for instance,

$$\hat{i}_b = (\sigma_e^2 g_n + \sigma_\mu^2 i_{n-1}) / (\sigma_e^2 + \sigma_\mu^2)$$

$$\hat{i} = (\sigma_e^2 g_n + \sigma_\mu^2 (i_{n-1} + i_{n+1})) / (\sigma_e^2 + 2\sigma_\mu^2)$$

The energies $\mathcal{E}(S2,S5)$ are the same as $\mathcal{E}(S1,S6)$ except calculated using the forward motion-compensated frame direction.

7.4.7.5   Estimating i_n, c

Once the state configuration is established at a pixel site, estimates for $I(\mathbf{x})$, $c(\mathbf{x})$ can be generated. This is done directly from the posterior. If $b(\mathbf{x}) = 1$, then $I(\mathbf{x})$ is interpolated from the previous and/or next frames (depending on $s(\mathbf{x})$, and c is set to a noise reduced version of the observed image. When $b(\mathbf{x}) = 0$, the image estimate is a temporal average (two or three frames) depending on occlusion, and $c(\mathbf{x})$ is interpolated spatially. Further computational savings can be had by exploiting the fact that all the energies involving g_n, I_n can be precomputed.

7.4.8   Pictures and comparisons

It is difficult to compare this system with previous work in noise reduction or blotch removal since this system treats the two problems jointly. Therefore, it makes compromises that sometimes give worse performance in one of the two domains, but overall the output pictures show good improvement. In some attempt to place this algorithm in context, a 256×256 subsection of the mobile and calendar sequence was corrupted with blotches that follow the prior for $c(\mathbf{x})$, and Gaussian noise of $\sigma_\mu^2 = 100$ was added. This is in keeping with the degradation model discussed here. A number of experiments were performed to evaluate the behavior as a blotch detector and as a noise reducer. The algorithm discussed above is called Joint Noise Reduction, Detection, and Interpolation (JONDI).

7.4.8.1   Blotch detection performance

As discussed previously, many of the currently available simple or fast blotch detection processes are based on the model in Equation (7.1), and all employ temporal motion-compensated frame differences, e.g., SDIa, SDIp, and ROD. JONDI uses the same information but includes, as well, spatial and textural smoothness.

Figure 7.15 shows a receiver operating characteristic (ROC) that compares the performance of JONDI with several detectors with respect to their blotch detection performance. Good performance is indicated by curves that are close to the top left corner of the graph. To create the characteristics, the processes were run with a range of parameter settings. In the case of SDIp, $T = 5{:}5{:}55$ (MATLAB notation), and the performance degrades as the threshold increases. For ROD, t_1 was varied $5{:}5{:}45$, with the other thresholds held at nominal values of $t_2 = 39.0$ and $t_3 = 55.0$.

The situation is more complicated with JONDI, Morris [33, 65], and JOMBANDI detectors [6, 41, 74] since there are two parameters to be set in each case. (JOMBANDI stands for Joint Model-Based Noise Reduction, Blotch Detection, and Interpolation.) However, from top right to bottom left, the points on the JONDI curve shown correspond to the following values for (Λ^c, Λ^b): (0.001,1.0), (0.1,1.0), (0.1,4.0), (0.1,8.0), (0.15,15). For JOMBANDI (see [41, 74]) the parameters having the same meaning were (0.15,1.0), (0.1,1.0), (0.15,4.0), (0.1,4.0). The lower curve (dashed) corresponds to these parameters with a 1-tap AR model (identical to Equation (7.1) except with a gain parameter in the prediction), whereas the upper curve (solid) corresponds to a 5-tap 3-DAR process. Ten iterations were used for JONDI, JOMBANDI, and Morris on each frame.

Like JONDI, the Morris detector has two parameters, Λ_o for occlusion smoothness and α for the occlusion penalty. The points on the ROC for this detector from top right to bottom left correspond to settings $(\Lambda_o, \alpha) = (4,1)$, (1,5), (4,5), (4,8), (4,10), (8,10), (8,20).

A multiresolution gradient-based block motion estimator was used to generate candidates for motion [6]. A block size of 9×9 pixels was employed with a 2-pixel overlap between blocks. The SDIp detector was used to initialize the $b(\mathbf{x})$ field for JONDI, using a threshold of 25 gray levels. The algorithm was run recursively; i.e., restored frames were used in the processing of the following frames.

The correct detection rate is measured as the fraction of pixels (out of the total number of missing sites) correctly set to 1 in $b(\mathbf{x})$. The false alarm rate is measured as the fraction of pixels incorrectly flagged as missing out of all the possible uncorrupted sites. Figure 7.15 shows clearly that JONDI outperforms SDIp, ROD, Morris, and JOMBANDI across a wide range of parameters. It is interesting to note that ROD and JOMBANDI (with a 1-tap

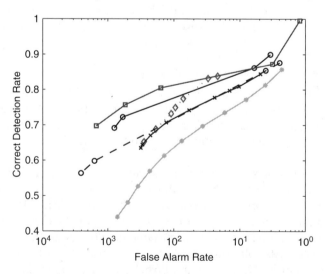

Figure 7.15 Performance of JONDI (•), JOMBANI (– ○ – [1 tap]; – ○ [5 tap]), Morris (◇), ROD (×), and SDIp (*) at $\sigma_\mu^2 = 100$.

AR process) perform almost the same at high false alarm rates. The Morris detector and JONDI have the same basic image sequence model in common, yet JONDI is able to continue giving reasonable performance at low false alarm rates (<.001) whereas Morris performance drops off sharply. This is because JONDI also incorporates priors on the blotches and seeks to improve the motion estimate as its iterations proceed. The 5-tap JOMBANDI gets close to the JONDI performance but does not reach it. JONDI is therefore gaining much improvement from the incorporation of the occlusion fields.

Figure 7.16 shows the result of JONDI on three frames from the corrupted sequence. The bottom row shows $(I(\mathbf{x}) - G(\mathbf{x})) + 128$ and so illustrates more clearly what has been removed from the dirty image. The combined blotch rejection and noise reduction features are clear. Also important is that the rotating ball is not damaged. Note, as well, that the corruption level in the test sequence is very high and, in fact, corruption at the same site in consecutive frames does occur.

7.4.8.2   Noise reduction performance

Figure 7.17 shows the dB improvement in signal to noise ratio (SNR) after processing with JONDI, the temporal Wiener filter [75], and the temporal recursive frame averaging filter [76]. To separate out the noise reduction component of JONDI from the missing data treatment component, the measurement of SNR was made only in those regions not corrupted by missing data. This does not, however, totally separate the two components since blotches can have an effect on processing for some distance *outside* their area.

The lowest curve shows the SNR of the degraded sequence at about 22 dB, and the top curve shows that JONDI at $\Lambda^c = 0.15$, $\Lambda^b = 4.0$ performs best. Changing Λ^b to 1.0 makes JONDI perform somewhere between the two temporal filters as far as noise reduction is concerned. This is sensible since a reduction in Λ^b implies that blotches are less convex, which is not the case. Note that as a noise reducer JONDI acts as a kind of automatically controlled recursive process. It should therefore perform similar to the algorithm outlined in [76]. The main difference for noise reduction is that JONDI incorporates a motion refinement step. Thus, initially, when there is little pathological motion in the sequence, JONDI does well. However, as the mobile in the scene starts to move rapidly, even the motion processes in JONDI begin to fail, and performance degrades to the normal recursive noise reducer level.

7.4.8.3   Real pictures

The top row of Figure 7.16 shows a zoom on part (300×300 pixels) of a typical real degraded sequence, showing damage with both missing data and noise. The motion is roughly upward. The damage in this case is caused by a tear at two parts in the frame. The next row shows a corresponding restoration for the last two frames with the algorithm described in this chapter. The restoration is good, and the noise reduction is effective despite the fact that there is no spatial constraint on the image data. Five iterations were used, and $\Lambda^b = 4.0, \Lambda^o = 2.0$, and $\Lambda^c = 0.15$. The same multiscale gradient-based

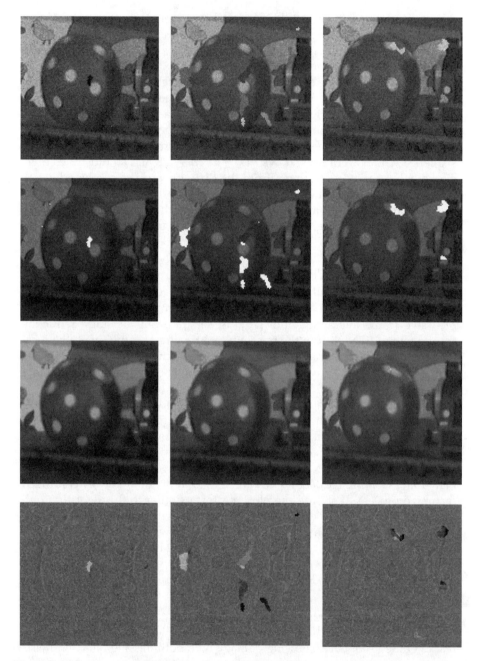

Figure 7.16 Section of mobile and calendar sequence frames 6,7,8. Top row: corrupted with blotches and $\sigma_\mu^2 = 100$; Second row: blotch detection $b(\mathbf{x})$ superimposed in bright white; Third row: algorithm result $\Lambda^c = 0.15$, $\Lambda^b = 4.0$; Bottom row: difference between images in top and third rows offset by 128.

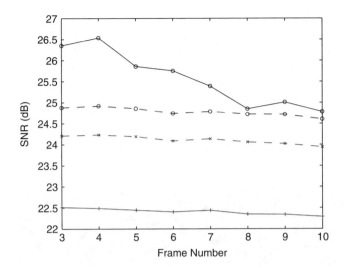

Figure 7.17 Noise reduction performance for $\sigma_\mu^2 = 100$. JONDI ($\circ$ –), temporal Wiener ($\times$ - -), and recursive filtering ($\circ$ - -). Algorithm settings are $\Lambda^c = 0.15$, $\Lambda^b = 4.0$ ($\circ$ -). Noisy sequence is represented by the bottom line.

motion estimator [6] used for the artificial case above was employed to initialize the motion field, and the block size was 17×17 pixels. For these results, $\sigma_\mu = 20.0$. This is a user-defined parameter since noise reduction is highly subjective. The next rows show the final configured detection, corruption, motion, and occlusion fields.

The fourth row shows that the motion is initially severely affected by the tear (in the region of the tear), but the algorithm corrects this, as shown on the right. The estimated corruption field is roughly equal to the degraded image where a blotch is detected and is very flat otherwise. This is sensible, given the GMRF used as a prior. The occlusion fields are able to prevent distortion in one frame from affecting the next, since the forward occlusion field is correctly set in frame 2 in the region that is damaged by the tear in frame 3. Previous work did not account for occlusion and distortion therefore tended to leak between frames. Note, however, that the occlusion indicators are linked only to intensity, not motion. This implies that they do not infer occlusion correctly but only to the extent that it helps the restoration.

It is interesting that even though the estimated pictures (second row) show a definite reduction in artifacts, especially noise, not all the damaged pixels appear to have been detected. See the top right corner of Figure 7.14. This is because the system is accounting for some of the low contrast missing data through the noise process and not through the blotch corruption process. This kind of interaction is under investigation. The same observation could be made about the pictures in Figure 7.16.

7.4.8.4 Color

Figure 7.18 shows a series of images from a color sequence that was processed with JONDI. The YUV color space was used. The motion manipulation, blotch detection, and noise reduction were performed only on the image luminance component, whereas the blotch reconstruction process was performed on all three planes. Noise reduction and blotch detection were restricted to luminance only because most of the signal energy lies in that component. It is true that sometimes blotches can occur in the color planes only (especially in the case of digital dropout), but this can be dealt with by

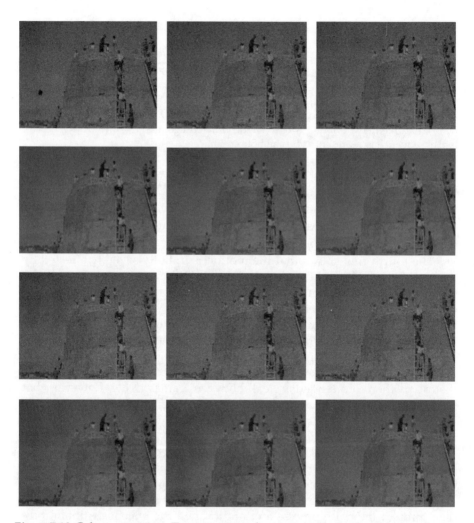

Figure 7.18 Color restoration: Top two rows show, respectively, three frames from a real archived sequence and the corresponding restoration. Bottom two rows show more images from the same sequence. Color versions of these images are available online at http://www.mee.tcd.ie/~ack.

performing the detection across all three planes. In this case limiting detection and noise reduction to the luminance component added the benefit of keeping computation low while still doing an effective job across the majority of degraded material. The identical parameter settings used in the previous example were also used here. The reduction in noise in particular was remarkable. Note also that a large dark blotch on the left of the first image was correctly removed. There is generally a lower level of corruption by blotches in real pictures as compared with the mobile/calendar experiment discussed above.

7.4.9 Relationships

The pixelwise relaxation approach that incorporates occlusion shows interesting links with previous works. These are considered next in chronological order.

Morris [33, 34, 65] considered that detection of blotches could be articulated by the detection of sites at which both forward and backward discontinuities occur. In the framework just presented, the Morris problem is then one of detecting when $O_f = 1$ and $O_b = 1$. To allow this in JONDI, the pixel states would have to be altered to consider only occlusion, such that each pixel state $s' = [O_b, O_f]$ instead of $[b, O_b, O_f]$. Then b is no longer a variable of interest. The clean data likelihood in Equation (7.8) can then be changed such that $I_n = G_n$, $I_{n-1} = G_{n-1}$, $I_{n+1} = G_{n+1}$. Then the problem would be to estimate s' by maximizing $p(S'|G_{n-1}, G_n, G_{n+1})$. The motion, detection, and corruption priors are then superfluous, since b, I are no longer variables of interest, and the same algorithm results. In essence, the Morris detector is detecting temporal discontinuities based on DFD just like the SDI detectors but also includes a prior to encourage spatial smoothness in the discontinuity output. This process relies heavily on reasonable motion estimation and is poor at motion edges. In 1997, Chong et al. [77] improved upon this idea by suppressing the flagging of occlusion at motion edges. A preprocess was required to configure a rough guess motion edge field, and this was used to configure the prior.

Kokaram et al. [6, 34, 41] proposed that the image sequence can be locally modeled by a 3-D spatio-temporal autoregressive process or 3-D linear predictor. This replaced the pixel difference image model in Equation (7.1) with a linear prediction error based on a spatio-temporal volume. There was no direction associated with this prediction, and this caused difficulties in occluded areas. Deleting the occlusion variable from JONDI, incorporating a linear prediction likelihood, and relinquishing the notion of a pixel state that is other than just luminance yielded JOMBANDI as presented in [41].

Wooi Boon et al. [55] solved the 3-DAR direction problem in a deterministic framework by choosing the direction that gave the lowest prediction error. In 1999, Van Roosmalen [35, 36] incorporated a direction indicator into a 3-DAR reconstruction framework. This is identical to using occlusion in the JONDI framework, if the likelihood is altered, and it is assumed that the detection field $b(\mathbf{x})$ and motion were given.

7.4.10 Why can JONDI work?

The unique aspect about JONDI, aside from the obvious joint treatment of several variables, is that it is connecting the picture-building process with the discontinuity detection and motion-estimation process. The energies that are used to select pixel states are trading off the luminance deviation from a linear estimate of the current pixel, $\hat{i}, \hat{i}_b, \hat{i}_f$, against the smoothness in the local variable field. In other words, if the local luminance is very far away from the luminance in nearby frames, and this behavior also occurs in nearby sites, then there must be some problem. Viewed in this light, detection using the ROD and filtering approaches in general can be seen as providing an alternative estimate for the current pixel. The ROD uses the idea that a rank-order filter on the local region is a reasonable estimate of the local true pixel intensity. Hence, deviations from that could flag a problem. It would be interesting to generalize this concept within JONDI by expressing the likelihood in a more general nonlinear fashion, but that is left for future work.

Furthermore, what JONDI explicitly makes clear is that the need for fully volumetric processing in image sequences is unnecessary if motion can be handled properly. In other words, most of the time each frame in the sequence can be synthesized by rearranging frames in the past or in the future. Therefore, if that rearrangement could be estimated in some robust way, a cut-and-paste operation would be sufficient to synthesize picture material. The need for a 3-D autoregressive process (for instance) is therefore restricted to cases in which the sequence behavior is of a higher order, for instance, a video texture like flowing water or light reflecting off the ocean. In general, the need for spatio-temporal reconstruction is motivated more, perhaps, by an attempt to recover from errors in motion estimation itself rather than sequence modeling.

Finally, it is interesting to note that the estimates for $\hat{i}$ in JONDI are all the optimal Wiener estimates given Gaussian noise. In fact, $\hat{i}$ is very similar indeed to the temporal noise reduction filter first presented by Katsagellos et al. [75] in 1989.

7.4.11 Computational comments

The main advantage of JONDI is that by adopting simple temporal models, the computation for the state estimation is very low per pixel site. Large parts of the energy expressions can be precomputed. This is in contrast to previous work that incorporated 3-DAR processes [6], which did give better robustness to small motion errors but at considerably more computational cost. In that process, the solution for the AR parameters alone required the inversion of at least a 5×5 matrix for *every* block of pixels. In JONDI, there is no such estimation step. Without the need for estimating the parameters of a stochastic process, the motion estimation load becomes dominant in JONDI and accounts for 80% of the time spent on one iteration. A single iteration (for both state and motion estimation) takes about 15 sec per 720×576 frame on a PIII 300 MHz PC (using nonoptimized code), and just 2 sec of this is due to the state estimation step. The surprising aspect about JONDI, as far as the ROC tests are concerned, however, is that it is able to perform better than other Bayesian detectors at a lower computational cost.

7.5 Failure in the motion model

Much of the work reviewed and presented thus far considers that it is possible to write an image sequence model of the kind in Equation (7.1). More correctly, it is expected that the use of the Gaussian error term $e(\mathbf{x})$ is good enough to account for deviations from the model. This is hardly ever the case. Fast motion of objects causes blurring, and many "interesting objects," e.g., clothing, are not rigid. This means that in some parts of any sequence it will be impossible to model the behavior. In this situation most of the current missing data techniques fail. Damage to fast-moving material is typically annoying to most viewers and no processing in those regions is often preferred.

In recent efforts, Bornard [63, 64] and Rares et al. [78, 79, 80, 81] have considered these problems from the practical standpoint of processing hours as opposed to seconds of video material. They have acted upon a well-known observation by the users of restoration tools. Whenever motion estimators fail, they tend to cause a high degree of false alarms in roughly the same location in consecutive frames. Because blotches should not occur in the same place in consecutive frames, this behavior indicates a problem with the image material. Detection of this phenomenon allows the blotch treatment process to be turned off before any damage can be done. By building a five-frame version of the Morris [65] detector, Bornard [63] is able to detect areas where discontinuities are flagged regularly over the five-frame aperture. Although the process is conservative in what it treats, initial results from Bornard's work [64] are very encouraging. A classification approach to this problem has been taken by Rares et al. [78, 81]. Rares presents a taxonomy of pathological motion in [78] and attempts to classify the regions showing discontinuities as pathological or nonpathological. This is a long-term approach as there are implications for video processing in general.

7.6 Final comments and possible futures

Missing data treatment in video and film archives has matured to the point that simple algorithms are currently appearing in industry standard hardware. Although the more complex Bayesian approaches have yet to be implemented in such systems, recent developments have shown that low-cost Bayesian inference is possible.

Coping with pathological motion remains an issue. While Rares and Bornard have set the groundwork for considering this issue, the integration of model failure into restoration tasks in general is still an open question. Allowing the state $O_f = 1, O_b = 1$ in the Bayesian framework presented here could be one solution to the integration issue. However, this may not work since that occlusion state and the state $b = 1$ would then be heavily correlated, at least in the context of the discussion given in this chapter. Further work in this direction is currently under way.

The reader may also notice that most of the work in missing data for degraded video and film has not considered the use of color information. The examples in Figures 7.1 and 7.2 illustrate that, in color sequences, black and white dirt against colored backgrounds is common. The reason for ignoring this powerful information is simply that the search is for generic tools that would operate equally well for black and white movies as well as different types of missing data problems, e.g., emulsion or water damage. In the latter case, color content is not necessarily as heavily contrasted. Nevertheless, it should be interesting to consider how color information could best be incorporated when it is advantageous.

It is important to note that not only archived footage suffers from dirt. Recent footage always contains some level of defects, and the availability of relatively low-cost software tools for assisting the clean-up process is increasingly in demand. As restoration systems continue to be deployed, either for real-time video broadcast [5] or film scanning and postproduction [82] for digital cinema or DVD, users will increasingly have more say in the requirements for good systems. This will ensure the need for further research and development in this area. It is encouraging that this field of signal and image processing, which for so long had been on the periphery of mainstream applications, is slowly but surely growing in prominence.

Acknowledgments

The author would like to thank the many members of the AURORA and BRAVA EU consortia who created an educational atmosphere over several years for this kind of work to arise. It is important, as well, to acknowledge the input from several archives and workers in archive houses, including the BBC (R&D unit: John Drewery among others), INA (Jean-Hugues Chenot, Raphael Bornard, and Louis Laborelli), and RTP (Joao Sequeira), both for material and for the understanding of how defects develop. It is clear that this kind of work cannot be well informed without communication with experienced archivists and signal-processing researchers working within archive houses.

References

[1] P. M. B. Van Roosmalen, A. Kokaram, and J. Biemond, "Noise reduction of image sequences as preprocessing for MPEG2 encoding," in *European Conference on Signal Processing (EUSIPCO '98)*, September 1998, vol. 4, pp. 2253–2256.

[2] F. Chen and D. Suter, "Motion estimation for noise reduction in historical films: MPEG encoding effects," in *6th Digital Image Computing: Techniques and Applications (DICTA 2002) Conference*, 2002, pp. 207–212.

[3] D. Suter and S. Richardson, "Historical film restoration and video coding," in *Picture Coding Symposium*, 1996, pp. 389–394.

[4] P. Schallauer, G. Thallinger, and M. J. Addis, "Diamant — digital film manipulation platform," in *IEE Seminar on Digital Restoration of Film and Video Archives*, January 2001, pp. 3/1–3/3.

[5] S. M. Sommerville, "Archangel – automated real-time archive restoration," in *IEE Seminar on Digital Restoration of Film and Video Archives*, January 2001, pp. 5/1–5/30.

[6] A. C. Kokaram, *Motion Picture Restoration: Digital Algorithms for Artefact Suppression in Degraded Motion Picture Film and Video*, Springer-Verlag, London, 1998.

[7] A. C. Kokaram, "Reconstruction of severely degraded image sequence," in *Image Analysis and Processing*, vol. 2, Springer-Verlag, Berlin, 1997.

[8] D. N. Sidorov and A. C. Kokaram, "Suppression of moire patterns via spectral analysis," in *SPIE Conference on Visual Communications and Image Processing*, January 2002, vol. 4671, pp. 895–906.

[9] D. N. Sidorov and A. C. Kokaram, "Removing moire from degraded video archive," in *XI European Signal Processing Conference (EUSIPCO 2002)*, September 2002.

[10] P. J. W. Rayner, S. Armstrong, and A. C. Kokaram, "Restoring video images taken from scratched 2-inch tape," in *Workshop on Non-Linear Model Based Image Analysis, NMBIA'98*, S. Marshall, N. Harvey, and D. Shah, eds., Springer-Verlag, Berlin, 1998, pp. 83–88.

[11] S. Marshall and N. R. Harvey, "Film and video archive restoration using mathematical morphology," in *IEE Seminar on Digital Restration of Film and Video Archives*, January 2001, pp. 9/1–9/5.

[12] S.-J. Ko, S.-H. Lee, S.-W. Jeon, and E.-S. Kang, "Fast digital image stabilizer based on gray-coated bit-plane matching," *IEEE Transactions on Consumer Electronics*, August 1999, vol. 45, no. 3, pp. 598–603.

[13] K. Uomori, A. Morimura, H. Ishii, T. Sakaguchi, and Y. Kitamura, "Automatic image stabilizing system by full-digital signal processing," *IEEE Transactions on Consumer Electronics*, August 1990, vol. 36, no. 3, pp. 510–519.

[14] J. Tucker and A. de Sam Lazaro, "Image stabilization for a camera on a moving platform," in *Proceedings of the IEEE Pacific Rim Conference on Communications, Computers, and Signal Processing*, May 1993, vol. 2, pp. 734–737.

[15] S.-J. Ko, S.-H. Lee, and K.-H. Lee, "Digital image stabilizing algorithms based on bit-plane matching," *IEEE Transactions on Consumer Electronics*, August 1998, vol. 44, no. 3, pp. 617–622.

[16] K. Ratakonda, "Real-time digital video stabilization for multi-media applications," in *Proceedings International Symposium on Circuits and Systems*, May 1998, vol. 4, pp. 69–72.

[17]   T. Vlachos, "Simple method for estimation of global motion parameters using sparse translational motion vector fields," *Electronics Letters*, January 1998, vol. 34, no.1, pp. 60–62.

[18]   J.-M. Odobez and P. Bouthemy, "Robust multiresolution estimation of parametric motion models," *Journal of Visual Communication and Image Representation*, 1995, vol. 6, pp. 348–365.

[19]   L. Hill and T. Vlachos, "On the estimation of global motion using phase correlation for broadcasting applications," in *Seventh International Conference on Image Processing and Its Applications*, July 1999, vol. 2, pp. 721–725.

[20]   L. Hill and T. Vlachos, "Global and local motion estimation using higher-order search," in *Fifth Meeting on Image Recognition and Understanding (Miru 2000)*, July 2000, vol. 1, pp. 18–21.

[21]   F. Dufaux and J. Konrad, "Efficient robust and fast global motion estimation for video coding," *IEEE Transactions on Image Processing*, 2000, vol. 9, pp. 497–501.

[22]   A. Smolic and J.-R. Ohm, "Robust global motion estimation using a simplified m-estimator approach," in *IEEE International Conference on Image Processing*, September 2000.

[23]   W. Qi and Y. Zhong, "New robust global motion estimation approach used in MPEG-4," *Journal of Tsinghua University Science and Technology*, 2001.

[24]   P. M. B. Van Roosmalen, R. L. Lagndijk, and J. Biemond, "Flicker reduction in old film sequences," in *Time-Varying Image Processing and Moving Object Recognition 4*, V. Cappellini, ed., Elsevier Science, 1997, pp. 9–17.

[25]   P. M. B. Van Roosmalen, R. L. Lagndijk, and J. Biemond, "Correction of intensity flicker in old film sequences," Submitted to *IEEE Transactions on Circuits and Systems for Video Technology*, December 1996.

[26]   T. Vlachos and G. A. Thomas, "Motion estimation for the correction of twin-lens telecine flicker," in *IEEE International Conference on Image Processing*, September 1996, vol. 1, pp. 109–112.

[27]   P. Read and M.-P. Meyer, *Restoration of Motion Picture Film*, Butterworth-Heinemann, London, 2000.

[28]   V. Bruni and D. Vitulano, Scratch detection via underdamped harmonic motion," *International Conference on Pattern Recognition*, 2002, pp. III: 887–890.

[29]   L. Joyeux, S. Boukir, and B. Besserer, "Film line scratch removal using Kalman filtering and Bayesian restoration," in *Fifth IEEE Workshop on Applications of Computer Vision*, 2000, pp. 8–13.

[30]   L. Joyeux, S. Boukir, B. Besserer, and O. Buisson, "Detection and removal of line scratches in motion picture films," in *IEEE International Conference on Computer Vision and Pattern Recognition*, June 1999, pp. 548–553.

[31]   L. Joyeux, S. Boukir, B. Besserer, and O. Buisson, "Reconstruction of degraded image sequences: Application to film restoration," *Image and Vision Computing*, 2001, no. 19, pp. 503–516.

[32]   A. C. Kokaram, "Detection and removal of line scratches in degraded motion picture sequences," in *Signal Processing VIII*, September 1996, vol. I, pp. 5–8.

[33]   A. C. Kokaram, R. Morris, W. Fitzgerald, and P. Rayner, "Detection of missing data in image sequences," *IEEE Image Processing*, November 1995, pp. 1496–1508.

[34]   A. C. Kokaram, R. Morris, W. Fitzgerald, and P. Rayner, "Interpolation of missing data in image sequences," *IEEE Image Processing*, November 1995, pp. 1509–1519.

[35]   P. M. B. Van Roosmalen, *Restoration of Archived Film and Video*, Ph.D. thesis, Technische Universiteit Delft, The Netherlands, September 1999.

[36]   P. M. B. Van Roosmalen, A. Kokaram, and J. Biemond, "Fast high quality interpolation of missing data in image sequences using a controlled pasting scheme," in *IEEE Conference on Acoustics Speech and Signal Processing (ICASSP'99)*, March 1999, vol. IMDSP 1.2, pp. 3105–3108.

[37]   G. R. Arce, "Multistage order statistic filters for image sequence processing," *IEEE Transactions on Signal Processing*, May 1991, vol. 39, pp. 1146–1161.

[38]   R. Storey, "Electronic detection and concealment of film dirt," *SMPTE Journal*, June 1985, pp. 642–647.

[39]   M. J. Nadenau and S. K. Mitra, "Blotch and scratch detection in image sequences based on rank ordered differences," in *Fifth International Workshop on Time-Varying Image Processing and Moving Object Recognition*, September 1996.

[40]   A. C. Kokaram, "Advances in the detection and reconstruction of blotches in archived film and video," in *IEE Seminar on Digital Restoration of Film and Video Archives*, January 2001, pp. 7/1–7/6.

[41]   A. C. Kokaram and S. J. Godsill, "MCMC for joint noise reduction and missing data treatment in degraded video," *IEEE Transactions on Signal Processing, Special Issue on MCMC*, February 2002, vol. 50, no. 2, pp. 189–205,.

[42]   R. Storey, "Electronic detection and concealment of film dirt," *UK Patent Specification No. 2139039*, 1984.

[43]   A. C. Kokaram, R. Bornard, A. Rares, D. Sidorov, J.-H. Chenot, L. Laborelli, and J. Biemond, "Robust and automatic digital restoration systems: Coping with reality," in *International Broadcasting Convention (IBC)*, 2002.

[44]   A. C. Kokaram, R. Bornard, A. Rares, D. Sidorov, J.-H. Chenot, L. Laborelli, and J. Biemond, "Automatic digital restoration systems: Coping with reality," To appear in *Journal of the Society of Motion Picture and Television Engineers*, 2003.

[45]   F. Paisan and A. Crise, "Restoration of signals degraded by impulsive noise by means of a low distortion, non-linear filter," *Signal Processing*, 1984, vol. 6, pp. 67–76.

[46]   E. Decenciere Ferrandiere, *Motion Picture Restoration Using Morphological Tools*, Kluwer Academic Publishers, Dordrecht, 1999.

[47]   E. Decenciere Ferrandiere and J. Serra, "Detection of local defects in old motion pictures," in *VII National Symposium on Pattern Recognition and Image Analysis*, April 1997, pp. 145–150.

[48]   E. Decenciere Ferrandiere, *Mathematical Morphology and Motion Picture Restoration*, John Wiley and Sons, New York, 2001.

[49]   O. Buisson, *Analyse de Sequences d'Images Haute Resolution, Application a la Restauration Numerique de Films Cinematographiques*, Ph.D. thesis, Universite de La Rochelle, France, December 1997.

[50]   E. Decenciere Ferrandiere, *Restauration Automatique de Films Anciens*, Ph.D. thesis, Ecole des Mines de Paris, France, December 1997.

[51]   O. Buisson, B. Besserer, S. Boukir, and F. Helt, "Deterioration detection for digital film restoration," *Proceedings of Conference on Computer Vision and Pattern Recognition*, June 1997, vol. 1, pp. 78–84.

[52]   L. Tenze, G. Ramponi, and S. Carrato, "Blotches correction and contrast enhancement for old film pictures," in *IEEE International Conference on Image Processing*, 2000, pp. TP06.05.

[53] L. Tenze, G. Ramponi, and S. Carrato, "Robust detection and correction of blotches in old films using spatio-temporal information," in *Proceedings of SPIE International Symposium of Electronic Imaging 2002*, January 2002.

[54] T. Saito, T. Komatsu, T. Ohuchi, and T. Seto, "Image processing for restoration of heavily corrupted old film sequences," in *International Conference on Pattern Recognition 2000*, 2000, vol. III, pp. 17–20.

[55] G. Wooi Boon, M. N. Chong, S. Kalra, and D. Krishnan, "Bidirectional 3D autoregressive model approach to motion picture restoration," in *IEEE International Conference on Acoustics and Signal Processing*, April 1996, pp. 2275–2278.

[56] S. Kalra, M. N. Chong, and D. Krishnan, "A new autoregressive (AR) model based algorithm for motion picture restoration," in *IEEE International Conference on Acoustics and Signal Processing*, April 1997, pp. 2557–2560.

[57] M. Bertalmio, G. Sapiro, V. Caselles, and C. Ballester, "Image inpainting," in *Proceedings SIGGRAPH*, 2000.

[58] M. Bertalmio, C. Ballester, V. Caselles, G. Shapiro, and J. Verdera, "Filling in by joint interpolation of vector fields and gray levels," *IEEE Transactions on Image Processing*, August 2001, vol. 10, no. 8.

[59] A. C. Kokaram, "Parametric texture synthesis for filling holes in pictures," in *IEEE International Conference on Image Processing 2002*, September 2002.

[60] A. C. Kokaram, "A statistical framework for picture reconstruction using AR models," in *European Conference on Computer Vision Workshop on Statistical Methods in Video Processing*, June 2002, pp. 73–78.

[61] A. A. Efros and T. K. Leung, "Texture synthesis by non-paramectric sampling," in *Proceedings of the IEEE International Conference on Computer Vision (ICCV)*, September 1999, vol. 2, pp. 1033–1038.

[62] A. A. Efros and W. T. Freeman, "Image quilting for texture synthesis and transfer," in *Proceedings SIGGRAPH*, 2001, pp. 341–346.

[63] R. Bornard, *Approaches Probabilistes Appliquees a la Restauration Numerique d'Archives Televisees*, Ph.D. thesis, Ecole Centrale Des Arts Et Manufactures, Ecole Centrale Paris, 2002.

[64] R. Bornard, E. Lecan, L. Laborelli, and J.-H. Chenot, "Missing data correction in still images and image sequences," in *ACM Multimedia*, December 2002.

[65] R. D. Morris, *Image Sequence Restoration Using Gibbs Distributions*, Ph.D. thesis, Cambridge University, England, 1995.

[66] A. C. Kokaram and S. Godsill, "Joint detection, interpolation, motion and parameter estimation for image sequences with missing data," in *IEEE International Conference on Image Processing*, October 1997, pp. 191–194.

[67] A. C. Kokaram, "Practical MCMC for missing data treatment in degraded video," in *European Conference on Computer Vision Workshop on Statistical Methods in Video Processing*, June 2002, pp. 85–90.

[68] A. C. Kokaram and P. Rayner, "A system for the removal of impulsive noise in image sequences," in *SPIE Visual Communications and Image Processing*, November 1992, pp. 322–331.

[69] R. D. Morris and W. J. Fitzgerald, "Detection and correction of speckle degradation in image sequences using a 3D Markov random field," in *Proceedings International Conference on Image Processing: Theory and Applications (IPTA '93)*, June 1993.

[70] J. J. O Ruanaidh and W. J. Fitzgerald, *Numerical Bayesian Methods Applied to Signal Processing*, Springer-Verlag, London, 1996.

[71] J. Konrad and E. Dubois, "Bayesian estimation of motion vector fields," *IEEE Transactions on Pattern Analysis and Machine Intelligence*, September 1992, vol. 14, no. 9.

[72] J. Besag, "On the statistical analysis of dirty pictures," *Journal of the Royal Statistical Society B*, 1986, vol. 48, pp. 259–302,.

[73] A. C. Kokaram, "On missing data treatment for degraded video and film archives: A survey and a new Bayesian approach," To appear in *IEEE Journal of Image Processing*, 2004.

[74] A. C. Kokaram and S. J. Godsill, "Joint noise reduction, motion estimation, missing data reconstruction and model parameter estimation for degraded motion pictures," in *SPIE International Conference on Bayesian Inference for Inverse Problems*, July 1998.

[75] A. Katsagellos, J. Driessen, S. Efstratiadis, and R. Lagendijk, "Spatio-temporal motion compensated noise filtering of image sequences," in *SPIE VCIP*, 1989, pp. 61–70.

[76] E. Dubois and S. Sabri, "Noise reduction in image sequences using motion compensated temporal filtering," *IEEE Transactions on Communications*, July 1984, vol. 32, pp. 826–831.

[77] M. N. Chong, P. Liu, W. B. Goh, and D. Krishnan, "A new spatio temporal MRF model for the detection of missing data in image sequences," in *IEEE International Conference on Acoustics and Signal Processing*, April 1997, vol. 4, pp. 2977–2980.

[78] J. Biemond, A. Rares, and M. J. T. Reinders, "Statistical analysis of pathological motion areas," in *IEE Seminar on Digital Restoration of Film and Video Archives*, January 2001.

[79] J. Biemond, A. Rares, and M. J. T. Reinders, "Complex event classification in degraded image sequences," in *Proceedings of ICIP 2001 (IEEE)*, October 2001.

[80] J. Biemond, A. Rares, and M. J. T. Reinders, "Image sequence restoration in the presence of pathological motion and severe artifacts," in *Proceedings of ICASSP 2002 (IEEE)*, May 2002.

[81] J. Biemond, A. Rares, and M. J. T. Reinders, "A spatio temporal image sequence restoration algorithm," in *Proceedings of ICIP 2002 (IEEE)*, September 2002.

[82] I. McLean and S. Witt, "Telecine noise reduction," in *IEE Seminar on Digital Restoration of Film and Video Archives*, January 2001, pp. 2/1–2/6.

chapter 8

Video summarization

Cuneyt M. Taskiran and Edward J. Delp

Contents

8.1 Introduction

Deriving compact representations of video sequences that are intuitive for users and that let them easily and quickly browse large collections of video data is fast becoming one of the most important topics in content-based video processing. Such representations, which we will collectively refer to as *video summaries*, rapidly provide the user with information about the content of the particular sequence being examined while preserving the essential message. The need for automatic methods for generating video summaries is fueled both from the user and production viewpoints. With the proliferation of personal video recorder devices and handheld cameras, many users generate many times more video footage than they can digest. On the other hand, in today's fast-paced news coverage, programs such as sports and news must be processed quickly for production or their value quickly decreases. Such time constraints and the increasing number of services being offered place a large burden on production companies to process, edit, and distribute video material as quickly as possible.

In this chapter we review the current state of the art in automatic video summarization. Summarization, be it for a video, audio, or text document, is a challenging and ill-defined task since it requires the processing system to make decisions based on high-level notions such as the semantic content and relative importance of the parts of the documents with respect to each other. Evaluating resulting summaries is also a problem since it is hard to derive quantitative measures of summary quality. We will examine some of the approaches that have been proposed to deal with these problems.

8.1.1   Types of video summaries

The goal of video summarization is to process video sequences that contain high redundancy and make them more exciting, interesting, valuable, and useful for users. The properties of a video summary depends on the application domain, the characteristics of the sequences to be summarized, and the purpose of the summary. Some of the purposes that a video summary might serve are listed below.

1. Intrigues the viewer to watch the whole video. Movie trailers, prepared by highly skilled editors and high budgets, are the best examples of this type.
2. Lets the user decide if the whole video is worth watching. Summaries of video programs that might be used in personal video recorders are examples of this category. The user may have watched the episode that was recorded or may have already watched similar content and so might not want to watch the program after seeing the summary.
3. Helps the user locate specific segments of interest. For example, in distance learning, students can skip parts of a lecture with which they are familiar and instead concentrate on new material.
4. Lets users judge if the video being examined is relevant to their queries. In content-based image database applications, it is customary to return the results of queries as thumbnail images, which can be judged at a glance for relevance to the queries. The same task is time consuming for video sequences since search results may contain many long sequences containing hundreds of shots. Presenting the summaries of the results would be much more helpful.

5. Enables users of pervasive devices, such as PDAs, palm computers, and cellular phones, to view video sequences, which these devices otherwise would not be able to handle due to their low processing power. Using summaries also results in significant downloading cost savings for such devices. An example of such an application is described in [1]. It uses annotations based on the MPEG-7 standard.

6. Gives the viewer all the important information contained in the video. These summaries are intended to replace watching the whole video. Executive summaries of long presentations would be an example of this type.

The above list, while not complete, illustrates the wide range of different types of video summaries one would like to generate. We observe that for most applications video summaries mainly serve two functions: the *indicative* function, in which the summary is used to indicate what topics of information are contained in the original program, and the *informative* function, in which the summary is used to cover the information in the source video as much as possible subject to the summary length. These summary applications are not independent, and video summarization applications often will be designed to achieve a mixture of them. Naturally, there is no single approach, neither manual nor automatic, that will apply to all types of video summaries.

8.1.2 Terminology

Before proceeding further, we will introduce the terminology that will be used throughout this chapter. We use the term *video summarization* to denote any method that can be used to derive a compact representation of a video sequence. Once a summary is obtained after processing the source video, it has to be presented to the user. We use the term *summary visualization*, or just *visualization*, to refer to the method that is used to display the summary. There are two main categories of summary visualizations:

1. In static visualization methods, a number of representative frames, often called keyframes, are selected from the source video sequence and are presented to the user, sometimes accompanied by additional information such as timestamps and closed caption text [2–8]. We refer to such visualizations as *video abstractions*. This type of summary visualization is sometimes referred to as the *storyboard presentation*.

2. Dynamic visualization methods generate a new, much shorter video sequence from the source video [9–14]. We refer to dynamic visualizations as *video skims*.

Unfortunately, there is little agreement in the literature on the terminology used to describe visualizations. For example, Hanjalic and Zhang define a video abstract as a "compact representation of a video sequence" [15], which denotes a general representation and corresponds to our definition of video summary. On the other hand, Lienhart [11] uses the same term to refer to what we defined to be video skims.

For video skims, the duration of the summary is generally specified by the user in terms of the summarization ratio (SR), which is defined as the ratio of the duration of the video skim to the duration of the source video. For video abstracts, the user may specify the number of keyframes to be displayed.

The chapter is organized as follows: In Section 8.2 we examine the numerous approaches proposed to automatically generate summaries from video programs. Methods used for summary visualization are investigated in Section 8.3. Evaluation of video summaries is examined in Section 8.4. Finally, some concluding remarks are given in Section 8.5.

8.2   Approaches to video summary generation

8.2.1   General approach to video summary generation

Most video content may be broadly categorized into two classes:

1. *Event-Based Content.* These types of video programs contain easily identifiable story units that form either a sequence of different events or a sequence of events and nonevents. Examples of the first kind of programs are talk shows and news programs where one event follows another and their boundaries are well defined. For talk shows each event contains a different guest, whereas for news programs each event is a different news story. The best example of programs in which a sequence of events and nonevents occur are sports programs. Here, the events may correspond to highlights such as touchdowns, home runs, and goals.
2. *Uniformly Informative Content.* These are programs that cannot be broken down to a series of events as easily as event-based content. For this type of content, all parts of the program may be equally important for the user. Examples of this type of content are sitcoms, presentation videos, documentaries, soap operas, and home movies.

Note that the distinction given above is not clear-cut. For example, for sitcoms one can define events according to the appearance of canned laughter. Movies form another example; most action movies have a clear sequence of action and nonaction segments.

For event-based content, since the types of events of interest are well defined, one can use knowledge-based event detection techniques. In this case, the processing will be domain specific and a new set of events and event detection rules must be derived for each application domain, which is a disadvantage. However, the summaries produced will be more reliable than those generated using general summarization algorithms. This type of algorithm is examined in Section 8.2.5.

If domain-based knowledge of events is not available or if one is dealing with uniformly informative content such as documentaries, one has to resort to more general summarization techniques. The main idea behind these techniques is to get rid of the redundancy in a video sequence by clustering similar parts of the sequence together. This may be done in two ways. In the top-down clustering approach, the source video is first divided into segments. One way to achieve this is to perform shot boundary detection on the sequence. Another way is to perform time-constrained clustering to identify segments. Once the segments are identified, they are clustered to collect similar segments together. In the bottom-up clustering approach, the segment detection step is skipped and frames from the video are clustered directly.

Once the content clusters in the video sequence are identified, each cluster is represented using either keyframes or portions of the segments belonging to clusters. Representation of clusters for summarization display purposes will be examined in depth in Section 8.3. In this section we investigate various approaches that have been proposed to generate video summaries.

8.2.2 Speedup of playback

A simple way to compactly display a video sequence is to present the complete sequence to the user but increase the playback speed. A technique known as time scale modification can be used to process the audio signal so that the speedup can be made with little distortion [16]. The compression allowed by this approach, however, is limited to a summarization ratio (SR) of 1.5–2.5, depending on the particular program genre [17]. Based on a comprehensive user study, Amir et al. report that for most program genres and for novice users, an SF of 1.7 can be achieved without significant loss in comprehension [16]. However, this SR value is not adequate for most summarization applications, which often require an SR of 10–20.

8.2.3 Techniques based on frame clustering

Dividing a video sequence into segments and extracting one or more keyframes from each segment has long been recognized as one of the simplest and most compact ways of representing video sequences. For a survey of keyframe extraction techniques, the reader is referred to [15]. These techniques generally focus on the image data stream only. Color histograms, because of their robustness, have generally been used as the features for clustering.

One of the earliest works in this area is by Yeung and Yeo [2] using time-constrained clustering of shots. Each shot is then labeled according to the cluster to which it belongs, and three types of events — dialogue, action, and other — are detected based on these labels. Representative frames from each event are then selected for the summary. The Video Manga system by Uchihashi et al. [7] clusters individual video frames using YUV color histograms. Iacob, Lagendijk, and Iacob [18] propose a similar technique. However, in their approach video frames are first divided into rectangles, whose sizes depend on the local structure, and YUV histograms are extracted from these rectangles. Ratakonda, Sezan, and Crinon [19] extract keyframes for summaries based on the area under the cumulative action curve within a shot, where the action between two frames is defined to be the absolute histogram difference between color histograms of the frames. They then cluster these keyframes into a hierarchical structure to generate a summary of the program.

Ferman and Tekalp [20] select keyframes from each shot using the fuzzy c-clustering algorithm, which is a variation of the k-means clustering method, based on alpha-trimmed average histograms extracted from frames. Cluster validity analysis is performed to automatically determine the optimal number of keyframes from each shot to be included in the summary. This summary may then be processed based on user preferences, such as the maximum number of keyframes to view, and cluster merging may be performed if there are too many keyframes in the original summary.

The approaches proposed in [15] and [21] both contain a two-stage clustering structure, which is very similar to the method used in [20], but instead of performing shot detection, segments are identified by clustering of video frames. Hanjalic and Zhang [15] use features and cluster validity analysis techniques that are similar to those in [20]. Farin, Effelsberg, and de With [21] propose a two-stage clustering technique based on luminance histograms extracted from each frame in the sequence. First, in an approach similar to time-constrained clustering, segments in the video sequence are located by minimizing segment inhomogeneity, which is defined as the sum of the distances of all frames within a segment to the mean feature vector of the segment. Then, the segments obtained are clustered using the Earth-Mover's distance [22]. Yahiaoui, Merialdo, and Huet [23] first cluster frames based on the L_1 distance between their color histograms using a procedure similar to k-means clustering. Then, a set of clusters is chosen to maximize the coverage over the source video sequence.

An application domain that poses unique challenges for summarization is home videos. Since home videos generally have no plot or summary and contain very little editing, if any, the editing patterns and high-level video structure, which offer a strong cue in the summarization of broadcast programs, are absent. However, home videos are inherently time-stamped during recording. Lienhart [11] proposes a summarization algorithm in which shots are clustered at four time resolutions using different thresholds for each level of resolution using frame time stamps. Very long shots, which are common in home videos, are shortened by using the heuristic that during important events audio is clearly audible over a longer period of time than less important content.

8.2.4 Techniques based on frame clustering by dimensionality reduction

These techniques perform a bottom-up clustering of the video frames selected at fixed intervals. A high-dimensional feature vector is extracted from each frame; this dimensionality is then reduced either by projecting the vectors to a much lower dimensional space [24, 25] or by using local approximations to high-dimensional trajectories [6, 8]. Finally, clustering of frames is performed in this lower dimensional space.

DeMenthon, Kobla, and Doerman [6] extract a 37-dimensional feature vector from each frame by considering a time coordinate together with the three coordinates of the largest blobs in four intervals for each luminance and chrominance channel. They then apply a curve-splitting algorithm to the trajectory of these feature vectors to segment the video sequence. A keyframe is extracted from each segment. Stefanidis et al. [8] propose a similar system; however, they split the three-dimensional trajectories of video objects instead of feature trajectories.

Gong and Liu [24] use singular value decomposition (SVD) to cluster frames evenly spaced in the video sequence. Each frame is initially represented using three-dimensional RGB histograms, which results in 1125-dimensional frame feature vectors. Then, SVD is performed on these vectors to reduce the dimensionality to 150 and clustering is performed in this space. Portions of shots from each cluster are selected for the summary. Cooper and Foote [25] sample the given video sequence at a rate of one frame per second and extract a color feature vector from each extracted frame. The cosine of the angle between feature vectors is taken to be the similarity measure between them and a nonnegative similarity matrix is formed between all pairs of frames. Nonnegative matrix factorization (NMF) [26] is used to reduce the dimensionality of the similarity matrix. NMF is a linear approximation similar to SVD, the difference being the fact that the basis vectors are nonnegative.

8.2.5 Techniques using domain knowledge

As discussed in the beginning of this section, if the application domain of the summarization algorithm is restricted to event-based content, it becomes possible to enhance summarization algorithms by exploiting domain-specific knowledge about events. Summarization of sports video has been the main application for such approaches. Sports programs lend themselves well to automatic summarization for a number of reasons. First, the interesting segments of a program occupy a small portion of the whole content; second, the broadcast value of a program falls off rapidly after the event, so the processing must be performed in near real time; third, compact representations of sports programs have a large potential audience; and, finally, often there are clear markers, such as cheering crowds, stopped games, and replays, that signify important events.

Summarization of soccer has received a large amount of attention recently (see [27] for a survey of work in soccer program summarization). Li, Pan, and Sezan [28] develop a general model for sports programs in which events are defined to be the actions in a program that are replayed by the broadcaster. The replay is often preceded by a close-up shot of the key players or the audience. They apply their approach to soccer videos in which they detect close-up shots by determining if the dominant color of the shot is close to that of the soccer field. Ekin and Tekalp [27] divide each keyframe of a soccer program into nine parts and use features based on the color content to classify shots into long, medium, and close-up shots. They also detect shots containing the referee and the penalty box. Goal detection is performed similarly to [28] by detecting close-up shots followed by a replay. Cabasson and Divakaran [29] detect audio peaks and a motion activity measure to detect exciting events in soccer programs. Based on the heuristic that the game generally stops after an exciting event, they search the program for sequences of high motion followed by very little motion. If an audio peak is detected near such a sequence, it is marked as an event and included in the summary.

Domain knowledge can be very helpful, even for uniformly informative content. For example, He et al. [12] have proposed algorithms based on heuristics about slide transitions and speaker pitch information to summarize presentation videos.

8.2.6 Techniques used on closed-captions or speech transcripts

For some types of programs, a large portion of the informational content is carried in the audio. News programs, presentation videos, documentaries, teleconferences, and instructional videos are some examples of such content. Using the spoken text to generate video summaries becomes a powerful approach for these types of sequences. Content text is readily available for most broadcast programs in the form of closed captions. For sequences such as presentations and instructional programs, in which this information is not available, speech recognition may be performed to obtain the speech transcript. Once the text corresponding to a video sequence is available, one can use methods of text summarization to obtain a text summary. The portions of the video corresponding to the selected text may then be concatenated to generate the video skim. Processing text also provides a high level of access to the semantic content of a sequence that is not possible to achieve using the image content only.

Agnihotri et al. [30] search the closed-caption text for cue words to generate summaries for talk shows. Cues such as "please welcome" and "when we come back," in addition to domain knowledge about program structure, are used to segment the programs into parts containing individual guests and commercial breaks. Keywords are then used to categorize the conversation with each guest into a number of predetermined classes such as *movie* or *music*. In their ANSES system, Pickering, Wong, and Rueger [31] use key entity detection to identify important keywords in closed-caption text. Working under the assumption that story boundaries always fall on shot boundaries, they perform shot detection followed by the merging of similar shots based on the similarity of words they contain. They then detect the nouns in text using a part-of-speech tagger and use lexical chains [32] to rank the sentences in each story. The highest scoring sentences are then used to summarize each news story.

An example of a technique that uses automatic speech recognition (ASR) is proposed by Taskiran et al. [9]. The usage of ASR makes their method applicable to cases in which the closed-caption text is not available, such as presentations or instructional videos. In their approach the video is first divided into segments at the pause boundaries. Then, each segment is assigned a score using term frequencies within segments. Using statistical text analysis, dominant word pairs are identified in the program and the scores of segments containing these pairs are increased. The segments with highest scores are selected for the summary.

8.2.7   Approaches using multiple information streams

Most current summarization techniques focus on processing one data stream, which is generally image data. However, multi-modal data fusion approaches, in which data from images, audio, and closed-caption text are combined, offer the possibility to greatly increase the quality of the video summaries produced. In this section we look at a few systems that incorporate features derived from multiple data streams.

The MoCA project [14], one of the earliest systems for video summarization, uses color and action content of shots, among other heuristics, to obtain trailers for feature films. The Informedia project constitutes a pioneer and one of the largest efforts in creating a large video database with search and browse capabilities. It uses integrated speech-recognition, image-processing, and natural language-processing techniques for the analysis of video data [13, 33]. Video segments with significant camera motion, and those showing people or a text caption, are given a higher score. Audio analysis includes detection of names in the speech transcript. Audio and video segments selected for summary are then merged while trying to maintain audio/video synchronicity.

Ma et al. [34] propose a generic user attention model by integrating a set of low-level features extracted from video. This model incorporates features based on camera and object motion, face detection, and audio. An attention value curve is obtained for a given video sequence using the model, and portions near the crests of this curve are deemed to be interesting events. Then heuristic rules are employed, based on pause and shot boundaries, and the SR is used to select portions of the video for the summary. Another model-based approach is the computable scene model proposed by Chang and Sundaram [35], which uses the rules of film making and experimental observations in the psychology of audition. Scenes are classified into four categories using audio and video features.

8.3   Summary visualization types

After the video summarization is obtained by using the techniques described in Section 8.2, it has to be displayed to the user in an intuitive and compact manner. Depending on the desired summary type and length, information from all detected video clusters may be used. Alternatively, importance scores may be assigned to each cluster using various combinations of visual, audio, textual, and other features extracted from the video sequence, and only portions or keyframes extracted from the clusters with the highest scores may be included in the summary. In this section we review some of the summarization display approaches that have been proposed. These methods mainly fall into two categories: video abstracts based on keyframes extracted from video and video skims in which portions of the source video are concatenated to form a much shorter video clip.

8.3.1   Static visualizations or video abstracts

The simplest static visualization method is to present one frame from each video segment, which may or may not correspond to an actual shot, in a storyboard fashion. The problems with this method are that all shots appear equally important to the user and the representation becomes impractically large for long videos.

One way to alleviate this problem is to rank the video segments and to display only the representative keyframes belonging to the segments with highest scores. In order to further reflect the relative scores of the segments, the keyframes may be sized according to the score of the segment. This approach has been used in [2] and [7], in which keyframes from segments are arranged in a "video poster" using a frame-packing algorithm. In their PanoramaExcerpts system, Taniguchi, Akutsu, and Tonomura [3] use panoramic icons, which are obtained by merging consecutive frames in a shot, in addition to keyframes. In the Informedia project, time-ordered keyframes, known as filmstrips, were used as video abstracts [36].

Although video abstracts are compact, since they do not preserve the time-evolving nature of video programs, they present fundamental drawbacks. They are somewhat unnatural and hard to grasp for nonexperts, especially if the video is complex. Most techniques just present keyframes to the user without any additional metadata, like keywords, which can make the meaning of keyframes ambiguous. Finally, static summaries are not suitable for instructional and presentation videos as well as teleconferences, in which most shots contain a talking head and most of the relevant information is found in the audio stream. These deficiencies are addressed by dynamic visualization methods.

8.3.2   Dynamic visualizations or video skims

In these methods the segments with the highest scores are selected from the source video and concatenated to generate a video skim. While selecting portions of the source video to be included in the video skim, care must be exercised to edit the video on long audio silences, which generally correspond to spoken sentence boundaries. This is due to the experimentally verified fact that users find it very annoying when audio segments in the video skim begin in mid-sentence [9, 12].

8.3.3   Other types of visualizations

There are also some video-browsing approaches that may be used to visualize video content compactly and, hence, may be considered a form of video summarization.

As part of the Informedia project, Wactlar [33] proposes video collages, which are rich representations that display video data along with related keyframes, maps, and chronological information in response to a user query. In their BMOVIES system, Vasconcelos and Lippman [4] use a Bayesian network to classify shots as action, close-up, crowd, or setting based on motion, skin tone, and texture features. The system generates a timeline that displays the evolution of the state of the semantic attributes throughout the sequence. Taskiran et al. [5] cluster keyframes extracted from shots using color, edge, and texture features and present them in a hierarchical fashion using a similarity pyramid. In the CueVideo system, Srinivasan et al. [37] provide a video browser with multiple synchronized views. This browser allows switching between different views, such as storyboards, salient animations, slide shows with fast or slow audio, and full video while preserving the corresponding point within the video between all different views. Ponceleon and Dieberger [38] propose a grid, which they call the movieDNA, whose cells indicate the presence or absence of a feature of interest in a particular video segment. When the user moves the mouse over a cell, a window shows a representative frame and other metadata about that particular cluster. A system to build a hierarchical representation of video content is discussed in Huang et al. [39] where audio, video, and text content are fused to obtain an index table for broadcast news.

8.4   Evaluation of video summaries

Since automatic video summarization is an emerging field, serious questions remain concerning the appropriate methodology in evaluating the quality of the generated summaries. Many researchers do not include any form of quantitative summary evaluations. Evaluation of the quality of automatically generated video summaries is a complicated task because it is difficult to derive objective quantitative measures for summary quality. In order to be able to measure the effectiveness of a video summarization algorithm, one first needs to define features that characterize a good summary, given the specific application domain being studied. As discussed in Section 8.1, summaries for different applications will have different sets of desirable attributes. Hence, the criteria to judge summary quality will be different for different application domains.

Automated text summarization dates back at least to Luhn's work at IBM in the 1950s [40], which makes it the most mature area of media summarization. We will apply the terminology developed for text summary evaluation to evaluation of video summaries. Methods for the evaluation of text summaries can be broadly classified into two categories: *intrinsic* and *extrinsic* evaluation methods [41, 42]. In intrinsic evaluation methods, the quality of the generated summaries is judged directly based on the analysis of summary. The criteria used may be user judgment of fluency of the summary, coverage of key ideas of the source material, or similarity to an "ideal" summary prepared by humans. On the other hand, in extrinsic methods, the summary is evaluated with respect to its impact on the performance for a specific information retrieval task. To the best of our knowledge, all summary evaluations in video summarization literature to date have been of the intrinsic type, except [13].

For event-based content such as a sports program, in which interesting events are clearly marked, summaries might be scored using the percentage of events from the original program that they contain. Two such evaluations are given in [27] and [20]. Ekin and Tekalp [27] give precision and recall values for goal, referee, and penalty box detection, which are important events in soccer games. In Ferman and Tekalp's study [20], the video summary was examined to determine, for each shot, the number of redundant or missing keyframes in the summary. For example, if the observer thought that an important object in a shot was important but no keyframe contained that object, this resulted in a missing keyframe. Although they serve as a form of quantitative summary quality measure, the event detection precision and recall values given in these studies do not reflect the quality of the summaries from the user's point of view.

For uniformly informative content, in which events may be harder to identify, different evaluation techniques have been proposed. He et al. [12] determined the coverage of summaries of key ideas from presentation videos by giving users a multiple-choice quiz about the source video before and after watching a video skim extracted from it. The quizzes consisted of questions prepared by the presentation speakers that were thought to reflect the key ideas of the presentation. The quality of the video skims was judged by the increase in quiz scores. A similar technique was used by Taskiran et al. [9] in evaluating video skims extracted from documentaries. The quiz method has some serious drawbacks: First, it was found that it does not differentiate between different summarization algorithms adequately [9], so it is not useful by itself to judge between different algorithms; second, it is not clear how quiz questions can be prepared in an objective manner, except, perhaps, by authors of presentations who are not usually available; and finally, the concept of a "key idea" in a video program is ambiguous and may depend on the viewers.

Another intrinsic evaluation method is to have users rate the skims based on subjective questions, e.g., "Was the summary useful?" and "Was the summary coherent?" Such surveys were used in [11, 12, 13].

In intrinsic evaluation of text, summaries created by experts are generally used [41]. Using a similar approach would be much more costly and time consuming for video sequences. Another scheme for evaluation may be to present the segments from the original program to many people and let them select the segments they think should be included in the summary, thereby generating a ground truth. This seems to be a promising approach, although agreement among human subjects becomes an issue for this scheme.

Extrinsic evaluation methods offer a more promising alternative to summary evaluation by concentrating on the effect of the summary on some task that is related to the goal of the summary. However, the use of such methods has been very rare in video summarization work. The only extrinsic evaluation method we are aware of is the one used by Christel et al. [13]. In this study video skims extracted from documentaries were judged based on the performance of users on two tasks: fact finding, in which users used the video skims to locate video segments that answered specific questions, and gisting, in which users matched video skims with representative text phrases and frames extracted from source video.

8.5 Conclusion

In this chapter we have reviewed the current state of the art in automatic generation of summaries for video programs. We saw that compared with early approaches, such as [2, 14], the field has matured and new and more powerful approaches for summary generation and visualization have been proposed. Nevertheless, this field is still a very fast-growing one, and there remain many open questions, some of them mentioned in this section.

As we saw in Section 8.2, many video summarization algorithms concentrate on gathering information from one data stream, such as images, audio, or closed captions. Systematic gathering of information from all of these streams and fusing them to generate summaries will greatly enhance the summary quality, as seen from the high-quality summaries produced by the Informedia project [13]. Analyzing and fusing data from different information systems while keeping their synchronicity is a challenging task.

More emphasis needs to be placed on visualization of summaries. We need to go beyond the approach of simple lists of keyframes. Instead, the keyframes should be enhanced by extra information, such as keywords obtained from closed captions and icons providing at-a-glance access to the important aspects of the summary segments. Such systems have recently begun to appear [38]. In order to develop effective summary visualizations, learning user preferences and access patterns is a must.

The problem of deriving good evaluation schemes for automatically generated video summaries is still complex and open. We feel that valuable clues to this problem can be obtained by studying the numerous approaches proposed in text summary evaluations. Large user studies are needed in this area to decide which family of algorithms performs best for a given program genre.

A factor that can influence evaluation results is the value of the summarization factor used to obtain the video summaries. The evaluation results of the same summarization system can be significantly different when it is used to generate summaries of different lengths [41]. Ideal summarization factors for different program genres need to be investigated.

References

[1]   Belle L. Tseng, Ching-Yung Lin, and John R. Smith. Video summarization and personalization for pervasive mobile devices. In *Proceedings of SPIE Conference on Storage and Retrieval for Media Databases 2002*, volume 4676, pages 359–370, San Jose, CA, 23–25 January 2002.

[2]   Minerva M. Yeung and Boon-Lock Yeo. Video visualization for compact presentation and fast browsing of pictorial content. *IEEE Transactions on Circuits and Systems for Video Technology*, 7(5): 771–785, October 1997.

[3]   Y. Taniguchi, A. Akutsu, and Y. Tonomura. Panorama excerpts: Extracting and packing panoramas for video browsing. In *Proceedings of the ACM Multimedia*, pages 427–436, Seattle, WA, 9–13 November 1997.

[4]   Nuno Vasconcelos and Andrew Lippman. Bayesian modeling of video editing and structure: Semantic features for video summarization and browsing. In *Proceedings of IEEE International Conference on Image Processing*, Chicago, IL, 4–7 October 1998.

[5]   Cuneyt Taskiran, JauYuen Chen, Charles A. Bouman, and Edward J. Delp. A compressed video database structured for active browsing and search. In *Proceedings of the IEEE International Conference on Image Processing*, Chicago, IL, 4–7 October 1998.

[6]   Daniel DeMenthon, Vikrant Kobla, and David Doerman. Video summarization by curve simplification, In *Proceedings of the ACM Multimedia Conference*, pages 211–218, Bristol, England, 12–16 September 1998.

[7]   Shingo Uchihashi, Jonathan Foote, Andreas Girgenson, and John Boreczky. Video manga: Generating semantically meaningful video summaries. In *Proceedings of ACM Multimedia '99*, pages 383–392, Orlando, FL, 30 October–5 November 1999.

[8]   A. Stefanidis, P. Partsinevelos, P. Agouris, and P. Doucette. Summarizing video datasets in the spatiotemporal domain. In *Proceedings of the International Workshop on Advanced Spatial Data Management*, Greenwich, England, 6–7 September 2000.

[9]   Cuneyt M. Taskiran, Arnon Amir, Dulce Ponceleon, and Edward J. Delp. Automated video summarization using speech transcripts. In *Proceedings of SPIE Conference on Storage and Retrieval for Media Databases 2002*, volume 4676, pages 371–382, San Jose, CA, 20–25 January 2002.

[10] JungHwan Oh and Kien A. Hua. An efficient technique for summarizing videos using visual contents. In *Proceedings of the IEEE International Conference on Multimedia and Expo*, New York, NY, 30 July–2 August 2000.

[11] Rainer Lienhart. Dynamic video summarization of home video. In *Proceedings of SPIE Conference on Storage and Retrieval for Media Databases 2000*, volume 3972, pages 378–389, San Jose, CA, January 2000.

[12] Liwei He, Elizabeth Sanocki, Anoop Gupta, and Jonathan Grudin. Auto-summarization of audio–video presentations. In *Proceedings of the 7th ACM International Multimedia Conference*, pages 289–298, Orlando, FL, 30 October–5 November 1999.

[13] Michael G. Christel, Micheal A. Smith, Roy Taylor, and David B. Winker. Evolving video skims into useful multimedia abstractions. In *Proceedings of the ACM Computer–Human Interface Conference*, pages 171–178, Los Angeles, CA, 18–23 April 1998.

[14] Silvia Pfeiffer, Rainer Lienhart, Stephan Fischer, and Wolfgang Effelsberg. Abstracting digital movies automatically. *Journal of Visual Communication and Image Representation*, 7(4): 345–353, December 1996.

[15] Alan Hanjalic and HongJiang Zhang. An integrated scheme for automated video abstraction based on unsupervised cluster-validity analysis. *IEEE Transactions on Circuits and Systems for Video Technology*, 9(8): 1280–1289, 1999.

[16] Arnon Amir, Dulce B. Ponceleon, Brian Blanchard, Dragutin Petkovic, Savitha Srinivasan, and G. Cohen. Using audio time scale modification for video browsing. In *Proceedings of the 33rd Annual Hawaii International Conference on System Sciences*, Maui, HI, 4–7 January 2000.

[17] Barry Arons. SpeechSkimmer: A system for interactively skimming recorded speech. *ACM Transactions on Computer Human Interaction*, 4(1): 3–38, 1997.

[18] S. M. Iacob, R. L. Lagendijk, and M. E. Iacob. Video abstraction based on asymmetric similarity values. In *Proceedings of SPIE Conference on Multimedia Storage and Archiving Systems IV*, volume 3846, pages 181–191, Boston, MA, September 1999.

[19] Krishna Ratakonda, Ibrahim M. Sezan, and Regis J. Crinon. Hierarchical video summarization. In *Proceedings of SPIE Conference Visual Communications and Image Processing*, volume 3653, pages 1531–1541, San Jose, CA, January 1999.

[20] A. M. Ferman and A. M. Tekalp. Two-stage hierarchical video summary extraction to match low-level user browsing preferences. *IEEE Transactions on Multimedia*, 5(2): 244–256, 2003.

[21] Dirk Farin, Wolfgang Effelsberg, and Peter H. N. de With. Robust clustering-based video-summarization with integration of domain-knowledge. In *Proceedings of the IEEE International Conference on Multimedia and Expo 2002*, pages 89–92, Lausanne, Switzerland, 26–29 August 2002.

[22] Yossi Rubner, Leonidas Guibas, and Carlo Tomasi. The earth mover's distance, multi-dimensional scaling, and color-based image retrieval. In *Proceedings of the ARPA Image Understanding Workshop*, pages 661–668, New Orleans, LA, May 1997.

[23] I. Yahiaoui, B. Merialdo, and B. Huet. Comparison of multi-episode video summarization algorithms. *EURASIP Journal on Applied Signal Processing*, 3(1): 48–55, 2003.

[24] Yihong Gong and Xin Liu. Video summarization using singular value decomposition. In *Proceedings of the IEEE Conference on Computer Vision and Pattern Recognition*, Hilton Head Island, SC, pages 174–180, 13–15 June 2000.

[25] Matthew Cooper and Jonathan Foote. Summarizing video using non-negative similarity matrix factorization. In *International Workshop on Multimedia Signal Processing*, St. Thomas, US Virgin Islands, 9–11 December 2002.

[26] Daniel D. Lee and H. Sebastian Seung. Algorithms for non-negative matrix factorization. In *Proceedings of the Conference on Advances in Neural Information Processing Systems*, pages 556–562, Denver, CO, 27 November–2 December 2000.

[27] Ahmet Ekin and A. Murat Tekalp. Automatic soccer video analysis and summarization. In *Proceedings of SPIE Conference on Storage and Retrieval for Media Databases 2003*, volume 5021, pages 339–350, Santa Clara, CA, 20–24 January 2003.

[28] Baoxin Li, Hao Pan, and Ibrahim Sezan. A general framework for sports video summarization with its application to soccer. In *Proceedings of IEEE International Conference on Acoustic, Speech and Signal Processing*, pages 169–172, Hong Kong, 6–10 April 2003.

[29] Romain Cabasson and Ajay Divakaran. Automatic extraction of soccer video highlights using a combination of motion and audio features. In *Proceedings of SPIE Conference on Storage and Retrieval for Media Databases 2003*, volume 5021, pages 272–276, Santa Clara, CA, 20–24 January 2003.

[30] Lalitha Agnihotri, Kavitha V. Devera, Thomas McGee, and Nevenka Dimitrova. Summarization of video programs based on closed captions, in *Proceedings of SPIE Conference on Storage and Retrieval for Media Databases 2001*, volume 4315, pages 599–607, San Jose, CA, January 2001.

[31] Marcus J. Pickering, Lawrence Wong, and Stefan M. Rueger. ANSES: Summarization of news video. In *Proceedings of the International Conference on Image and Video Retrieval*, volume LNCS 2728, pages 425–434, Urbana, IL, 24–25 July 2003.

[32] Regina Barzilay and Michael Elhadad. Using lexical chains for text summarization. In *Proceedings of the Intelligent Scalable Text Summarization Workshop*, Madrid, Spain, July 1997.

[33] Howard D. Wactlar. Informedia — search and summarization in the video medium. In *Proceedings of the IMAGINA 2000 Conference*, Monaco, 31 January–5 February 2000.

[34] Yu-Fei Ma, Lie Lu, Hong-Jiang Zhang, and Mingjing Li. A user attention model for video summarization. In *Proceedings of ACM Multimedia '02*, pages 533–542, Juans Les Pins, France, 1–6 December 2002.

[35] Shih-Fu Chang and Hari Sundaram. Structural and semantic analysis of video. In *Proceedings of the IEEE International Conference on Multimedia and Expo*, New York, NY, 30 July–2 August 2000.

[36] Michael Christel, Alexander G. Hauptman, Adrienne S. Warmack, and Scott A. Crosby. Adjustable filmstrips and skims as abstractions for a digital video library. In *Proceedings of the IEEE Conference on Advances in Digital Libraries*, Baltimore, MD, 19–21 May 1999.

[37] S. Srinivasan, D. Ponceleon, A. Amir, B. Blanchard, and D. Petkovic. Engineering the web for multimedia. In *Proceedings of the Web Engineering Workshop*, WWW-9, Amsterdam, The Netherlands, 15–19 May 2000.

[38] Dulce Ponceleon and Andreas Dieberger. Hierachical brushing in a collection of video data. In *Proceedings of the 34th Hawaii International Conference on System Sciences*, Maui, HI, 3–6 January 2001.

[39] Qian Huang, Zhu Liu, Aaron Rosenberg, David Gibbon, and Behzad Shahraray. Automated generation of news content hierarchy by integrating audio, video, and text information. In *Proceedings of the IEEE International Conference on Acoustics, Speech, and Signal Processing*, pages 3025–3028, Phoenix, AZ, 15–19 March 1999.

[40] P. H. Luhn. Automatic creation of literature abstracts. *IBM Journal*, 2(2): 159–165, 1958.

[41] Hongyan Jing, Regina Barzilay, Kathleen McKeown, and Michael Elhadad. Summarization evaluation methods: Experiments and analysis. In *Proceedings of the AAAI Symposium on Intelligent Summarization*, Palo Alto, CA, 23–25 March 1998.

[42] I. Mani, D. House, G. Klein, L. Hirschman, L. Obrst, T. Firmin, M. Chrzanowski, and B. Sundheim. The TIPSTER SUMMAC text summarization evaluation. *National Institute of Standards and Technology*, October 1998.

chapter 9

High-resolution images from a sequence of low-resolution observations

Luis D. Alvarez, Rafael Molina, and Aggelos K. Katsaggelos[1]

Contents

9.1   Introduction

This chapter discusses the problem of obtaining a high-resolution (HR) image or sequences of HR images from a set of low-resolution (LR) observations. This problem has also been referred to in the literature by the names

[1] The work of L. D. Alvarez and R. Molina was partially supported by the Comisión Nacional de Ciencia y Tecnología under contract TIC2000-1275. The work of A. K. Katsaggelos was supported by the Motorola Center for Communications, Northwestern University.

of *super-resolution* (SR) and *resolution enhancement* (we will be using all terms interchangeably). These LR images are undersampled, and they are acquired either by multiple sensors imaging a single scene or by a single sensor imaging the scene over a period of time. For static scenes the LR observations are related by global subpixel shifts yet for dynamic scenes they are related by local subpixel shifts, due to object motion (camera motion, such as panning and zooming, can also be included in this model). In this chapter we will be using the terms *image(s)*, *image frame(s)*, and *image sequence frame(s)* interchangeably.

This is a problem encountered in a plethora of applications. Images and video of higher and higher resolution are required, for example, in scientific (medical, space exploration, surveillance) and commercial (entertainment, high-definition television) applications. One of the early applications of high-resolution imaging was with Landsat imagery. The orbiting satellite would go over the same area every 18 days, acquiring misregistered images. Appropriately combining these LR images produced HR images of the scene.

Increasing the resolution of the imaging sensor is clearly one way to increase the resolution of the acquired images. This solution, however, may not be feasible due to the increased associated cost and the fact that the shot noise increases during acquisition as the pixel size becomes smaller. On the other hand, increasing the chip size to accommodate the larger number of pixels reduces the data transfer rate. Therefore, signal-processing techniques, like the ones reviewed in this chapter, provide a clear alternative.

A wealth of research considers modeling the acquisition of the LR images and providing solutions to the HR problem. Literature reviews are provided in [1] and [2]. Work traditionally addresses the resolution enhancement of image frames that are filtered (blurred) and downsampled during acquisition and corrupted by additive noise during transmission and storage. More recent work, however, addresses the HR problem when the available LR images are compressed using any of the numerous image and video compression standards or a proprietary technique. In this chapter we address both cases of uncompressed and compressed data.

To illustrate the HR problem, consider Figures 9.1 and 9.2. In Figure 9.1(a), four LR images of the same scene are shown. The LR images are undersampled, and they are related by global subpixel shifts, which are assumed to be known. In Figure 9.1(b), the HR image obtained by bilinearly interpolating one of the LR images is shown (the upper left image of Figure 9.1(a) was used). In Figure 9.1(c), the HR image obtained by one of the algorithms described later in the chapter by combining the four LR images is shown. As can be seen, considerable improvement can be obtained by combining the information contained in all four images. In Figures 9.2(a) and (b), three consecutive HR frames and the corresponding LR frames of a video sequence are shown, respectively. The LR frames resulted by blurring, downsampling, and compressing the HR frames using the MPEG-4 standard at 128 Kbps. In Figure 9.2(c), the bilinearly interpolated frames corresponding to the LR frames in Figure 9.2(b) are shown, whereas in Figure

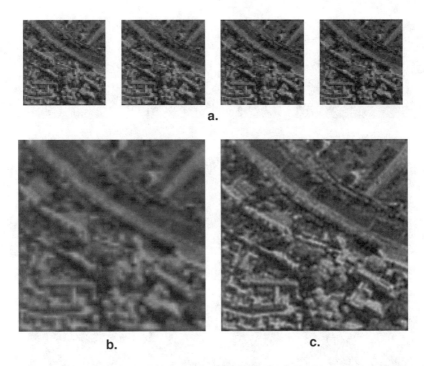

Figure 9.1 (a) Observed LR images with global subpixel displacement among them; (b) bilinearly interpolated image using the upper left LR image; (c) HR image obtained by combining the information in the LR observations using the algorithm in [59].

9.2(d), the HR frames obtained by one of the algorithms described later in the chapter are shown. Again, a considerable benefit results by using the available information appropriately.

An important question is what makes HR image reconstruction possible. We address this question with the aid of Figure 9.3. In it the grid locations of four LR images are shown. Each of these images represents a subsampled (aliased) version of the original scene. The four images, however, are related by subpixel shifts, as indicated in Figure 9.3. Each LR image therefore contains complementary information. With exact knowledge of shifts, the four LR images can be combined to generate the HR image shown on the right-hand side of Figure 9.3. If we assume that the resolution of this HR image is such that the Nyquist sampling criterion is satisfied, this HR image then represents an accurate representation of the original (continuous) scene. Aliased information can therefore be combined to obtain an alias-free image.

This example shows the simplest scenario and the ideal situation. However, there are several problems that should be taken into account, namely, the fact that the displacements between images may not be known in a realistic application and the motion may not be global but local instead. Furthermore, there are the blur and downsampling processes that have to be taken into account, together with the noise involved in the observation

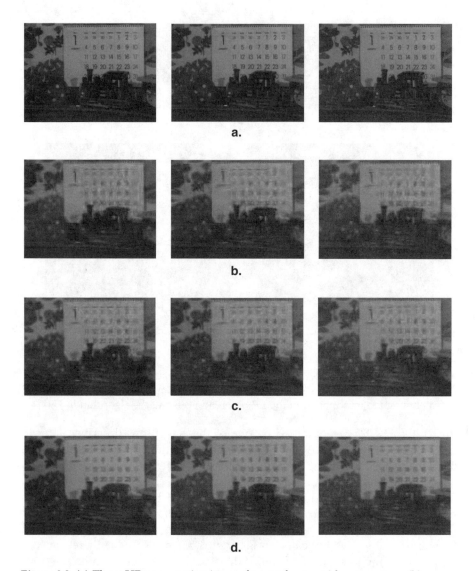

Figure 9.2 (a) Three HR consecutive image frames from a video sequence; (b) corresponding LR observations after blurring, subsampling, and compression using MPEG-4 at 128 Kbps; (c) bilinearly interpolated frames; (d) HR frames using the algorithm in [77].

process. This noise will be even more complicated when the images (or sequences) are compressed.

In this chapter we overview the methods proposed in the literature to obtain HR still images and image sequences from LR sequences. In conducting this overview, we develop and present all techniques within the Bayesian framework. This adds consistency to the presentation and facilitates comparison between the different methods.

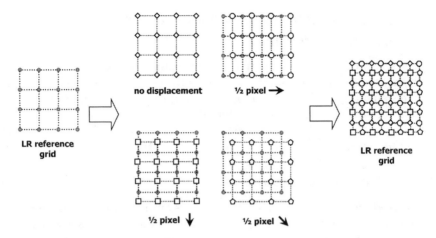

Figure 9.3   After taking several images of the same scene with subpixel displacement among them (several images from the same camera or one image from several cameras) or after recording a scene with moving objects in it, we can combine the observations to improve resolution.

The concept of an LR sequence will be understood in the broad sense to include the cases of global motion between images in the sequence and different motions of the objects in the image. We will also examine the case when the LR images have additionally been compressed. It is interesting to note that not much work has been reported on LR compressed observations with global motion between the observed images.

The HR techniques we discuss in this chapter do not directly consider the problem of HR video sequence reconstruction but instead concentrate on obtaining an HR still image from a short LR image sequence segment. All of these techniques may, however, be applied to video sequences by using a "sliding window" for processing frames. For a given high-resolution frame, a sliding window determines the set of low-resolution frames to be processed to produce the output.

The rest of this chapter is organized as follows. In Section 9.2 the process to obtain the low-resolution observations from the high-resolution image is described. In Section 9.3 the regularization terms used in high-resolution problems are presented; these regularization terms will include, in some cases, only the high resolution image, whereas in others they will include information about the motion vectors as well. In Section 9.4 the methods proposed to estimate the high-resolution images together with the high-resolution motion vectors will be presented. Both regularization and the observation process depend on unknown parameters; their estimation will be discussed in Section 9.5. Finally, in Section 9.6 we will discuss new approaches to the high-resolution problem.

9.2   Obtaining low-resolution observations from high-resolution images

In this section we describe the model for obtaining the observed LR images from the targeted HR image. We include both cases of a static image and an image frame from a sequence of images capturing a dynamic scene. As we have discussed in the previous section, the LR sequence may or may not be compressed. We first introduce the uncompressed case and then extend the model to include compression. In this section notation is also introduced.

9.2.1   Uncompressed observations

The pictorial depiction of the acquisition system is shown in Figure 9.4, in which the LR time-varying scene is captured by the camera (sensor). Let us denote the underlying HR image in the image plane coordinate system by $f(x,y,t)$ where t denotes temporal location and x and y represent the spatial coordinates. The size of these HR images is $PM \times PN$, where $P > 1$ is the magnification factor.

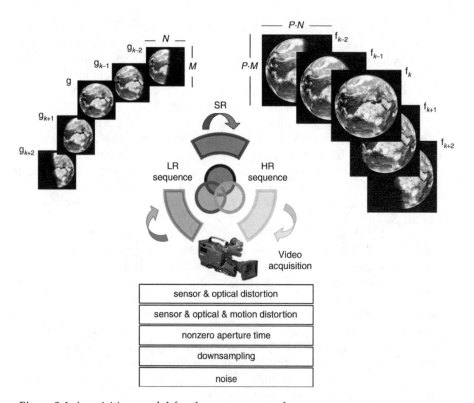

Figure 9.4 Acquisition model for the uncompressed case.

Using matrix–vector notation, the $PM \times PN$ images can be transformed into a $PM \times PN$ column vector, obtained by lexicographically ordering the image by rows. The $(PM \times PN) \times 1$ vector that represents the lth image in the HR sequence will be denoted by $\mathbf{f}_l$ with $l = 1, \ldots, L$. The HR image we seek to estimate will be denoted by $\mathbf{f}_k$.

Frames within the HR sequence are related through time. Here we assume that the camera captures the images in a fast succession, so we write

$$f_l(a,b) = f_k\left(a + d_{l,k}^x(a,b), b + d_{l,k}^y(a,b)\right) + n_{l,k}(a,b), \qquad (9.1)$$

where $d_{l,k}^x(a,b)$ and $d_{l,k}^y(a,b)$ denote the horizontal and vertical components of the displacement, that is, $d_{l,k}(a,b) = \left(d_{l,k}^x(a,b), d_{l,k}^y(a,b)\right)$, and $n_{l,k}(a,b)$ is the noise introduced by the motion compensation process. The above equation relates an HR gray level pixel value at location (a,b) at time l to the gray level pixel value of its corresponding pixel in the HR image we want to estimate, $\mathbf{f}_k$. For an image model that takes into account occlusion problems, see [3].

We can rewrite Equation (9.1) in matrix–vector notation as

$$\mathbf{f}_l = \mathbf{C}(\mathbf{d}_{l,k})\mathbf{f}_k + \mathbf{n}_{l,k}, \qquad (9.2)$$

where $\mathbf{C}(\mathbf{d}_{l,k})$ is the $(PM \times PN) \times (PM \times PN)$ matrix that maps frame $\mathbf{f}_l$ to frame $\mathbf{f}_k$, and $\mathbf{n}_{l,k}$ is the registration noise.

In Equation (9.2) we have related the HR images; now we have to relate the LR images to their corresponding HR ones, since we will have only LR observations. Note that the LR sequence results from the HR sequence through filtering and sampling. Camera filtering and downsampling the HR sequence produce, at the output, the LR sequence. This LR discrete sequence

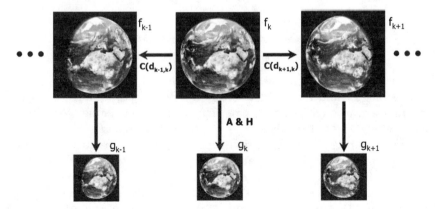

Figure 9.5 Relationship between LR and HR images. Note the use of motion compensation together with blurring and downsampling.

will be denoted by $g(i,j,t)$, where i and j are integer numbers that indicate spatial location and t is an integer time index. The size of the LR images is $M \times N$. Using matrix–vector notation, each LR image will be denoted by $\mathbf{g}_l$, where l is the time index of the image vector with size $(M \times N) \times 1$. The LR image $\mathbf{g}_l$ is related to the HR image $\mathbf{f}_l$ by

$$\mathbf{g}_l = \mathbf{AHf}_l \qquad l = 1, \dots, L, \tag{9.3}$$

where matrix $\mathbf{H}$ of size $(PM \times PN) \times (PM \times PN)$ describes the filtering of the HR image, and $\mathbf{A}$ is the downsampling matrix with size $MN \times (PM \times PN)$. We assume here for simplicity that all the blurring matrices $\mathbf{H}$ are the same, although they can be time dependent. The matrices $\mathbf{A}$ and $\mathbf{H}$ are assumed to be known. We also assume that there is no noise in Equation (9.3). Either we can add a new term that represents this noise or we can model it later when combining this equation with Equation (9.2).

Equation (9.3) expresses the relationship between the low and HR frames $\mathbf{g}_l$ and $\mathbf{f}_l$, whereas Equation (9.2) expresses the relationship between frames l and k in the HR sequence. Using these equations, we can now obtain the following equation that describes the acquisition system for an LR image $\mathbf{g}_l$ from the HR image $\mathbf{f}_k$ that we want to estimate:

$$\mathbf{g}_l = \mathbf{AHC}(\mathbf{d}_{l,k})\mathbf{f}_k + \mathbf{e}_{l,k}, \tag{9.4}$$

where $\mathbf{e}_{l,k}$ is the acquisition noise. This process is depicted pictorially in Figure (9.5).

If we assume that the noise during the acquisition process is Gaussian with zero mean and variance σ^2, denoted by $N(0, \sigma^2 I)$, the above equation gives rise to

$$P(\mathbf{g}_l \mid \mathbf{f}_k, \mathbf{d}_{l,k}) \propto \exp\left[-\frac{1}{2\sigma^2} \left\| \mathbf{g}_l - \mathbf{AHC}(\mathbf{d}_{l,k})\mathbf{f}_k \right\|^2 \right]. \tag{9.5}$$

Note that the above equation shows the explicit dependency of $\mathbf{g}_l$ on both the HR image $\mathbf{f}_k$ and the motion vectors $\mathbf{d}_{l,k}$. Both image and motion vector are unknown in most HR problems. This observation model was used by Hardie et al. [4], Elad and Feuer [5], Nguyen et al. [6], and Irani and Peleg [7], among others. The acquisition model of the HR frequency methods initially proposed by [8] can be also understood by using this model (see [9] for an excellent review of frequency-based super-resolution methods).

A slightly different model is proposed by Stark and Oskoui [10] and Tekalp et al. [11] (see also [12] and [13]). The observation model used by these authors is oriented toward the use of the projections onto convex sets (POCS) method in HR problems. In one case, this model results as the limit of Equation (9.5) when $\sigma = 0$. In other cases, the model imposes constraints on the maximum value of the difference between each component of $\mathbf{g}_l$ and $\mathbf{AHC}(\mathbf{d}_{l,k})\mathbf{f}_k$. Note that this corresponds to uniform noise modeling. We will encounter these noise models again in the compressed case.

9.2.2 Compressed observations

When the LR images have also been compressed, we have to relate them to the HR image we want to estimate, $\mathbf{f}_k$. The new scenario is shown in Figure 9.6.

When we have the compressed LR sequence, the uncompressed LR sequence is no longer available, so we cannot use $\mathbf{g}_l$, $l = 1, \dots, L$. These images are, however, used as an intermediate step since the compressed LR sequence results from the compression process applied to the LR sequence.

Let us now briefly describe a hybrid motion-compensated video compression process. The LR frames are compressed with a video compression system resulting in $y(i,j,t)$, where i and j are integer numbers that indicate spatial location and t is a time index. The size of the LR compressed images is $M \times N$. Using matrix–vector notation, each LR compressed image will be denoted by $\mathbf{y}_l$, where l is the time index and the image size is $(M \times N) \times 1$. The compression system also provides the motion vectors $v(i,j,l,m)$ that predict pixel $y(i,j,l)$ from some previously coded $\mathbf{y}_m$. These motion vectors that predict

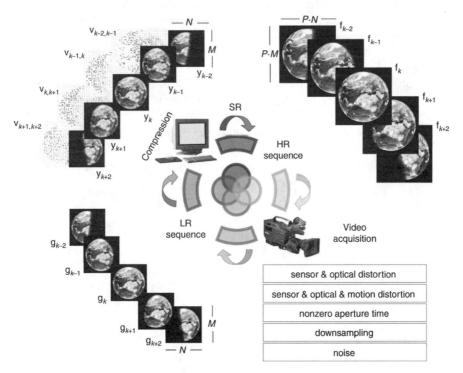

Figure 9.6 Acquisition model for the compressed problem.

$\mathbf{y}_l$ from $\mathbf{y}_m$ are represented by the $(2 \times M \times N) \times 1$ vector $\mathbf{v}_{l,m}$ that is formed by stacking the transmitted horizontal and vertical offsets.

During compression, frames are divided into blocks that are encoded with one of two available methods, intracoding or intercoding. For the first one, a linear transform such as the discrete cosine transform (DCT) is applied to the block (usually of size 8×8). The operator decorrelates the intensity data and the resulting transform coefficients are independently quantized and transmitted to the decoder. For the second method, predictions for the blocks are first generated by motion compensating previously transmitted image frames. The compensation is controlled by motion vectors that define the spatial and temporal offset between the current block and its prediction. Computing the prediction error, transforming it with a linear transform, quantizing the transform coefficients, and transmitting the quantized information refine the prediction.

Using all this information, the relationship between the acquired LR frame and its compressed observation becomes

$$\mathbf{y}_l = T^{-1}Q\left[T\left(\mathbf{g}_l - MC_l(\mathbf{y}_l^P, \mathbf{v}_l)\right)\right] + MC_l(\mathbf{y}_l^P, \mathbf{v}_l) \qquad l = 1, \ldots, L, \qquad (9.6)$$

where $Q[.]$ represents the quantization procedure, T and T^{-1} are the forward- and inverse-transform operations, respectively, and $MC_l(\mathbf{y}_l^P, \mathbf{v}_l)$ is the motion-compensated prediction of $\mathbf{g}_l$, formed by motion compensating the appropriate previously decoded frame(s), depending on whether the current frame at l is an I, P, or B frame (see [14]). This process is pictorially depicted in Figure 9.7.

Note that, to be precise, we should make clear that MC_l depends on $\mathbf{v}_l$ and only a subset of $\mathbf{y}_1, \ldots, \mathbf{y}_L$; however, we will keep the above notation for simplicity and generality.

We now have to rewrite Equation (9.6) using the HR image we want to estimate. The simplest way is to write

$$\mathbf{y}_l = T^{-1}Q\left[T\left(\mathbf{g}_l - MC_l(\mathbf{y}_l^P, \mathbf{v}_l)\right)\right] + MC_l(\mathbf{y}_l^P, \mathbf{v}_l)$$

$$\approx \mathbf{g}_l \text{ and using Eq. (9.4)}$$

$$= \mathbf{AHC}(\mathbf{d}_{l,k})\mathbf{f}_k + \mathbf{e}_{l,k}.$$

If we assume that the noise term $\mathbf{e}_{l,k}$ is $N(0, \sigma_l^2 I)$ (see [15] and references therein), we have

$$P(\mathbf{y}_l \mid \mathbf{f}_k, \mathbf{d}_{l,k}) \propto \exp\left[-\frac{1}{2\sigma_l^2}\left\|\mathbf{y}_l - \mathbf{AHC}(\mathbf{d}_{l,k})\mathbf{f}_k\right\|^2\right]. \qquad (9.7)$$

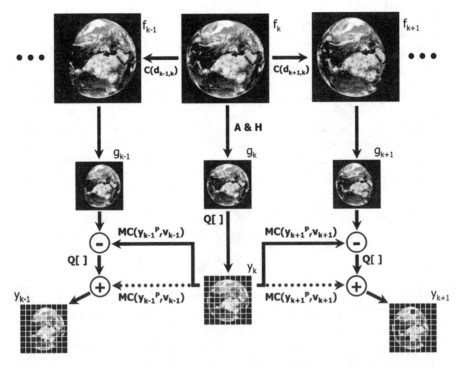

Figure 9.7 Relationship between compressed LR and HR images. Note the use of motion compensation together with blurring, downsampling, and quantization.

Two additional models for the quantization noise have appeared in the literature. The first one, called the *quantization constraint* (see Altunbasak et al. [16, 17] and also Segall et al. [18]), has associated probability distribution given by

$$P_{QC}(\mathbf{y}_l \mid \mathbf{f}_k, \mathbf{d}_{l,k}) = \begin{cases} constant \; if -\dfrac{q(i)}{2} \le \left[\mathbf{T}\big(\mathbf{AHC}(\mathbf{d}_{l,k})\mathbf{f}_k - MC_l(\mathbf{y}_l^P, \mathbf{v}_l)\big)\right](i) \le \dfrac{q(i)}{2}, \forall i \\ 0 \; elsewhere \end{cases},$$

$$(9.8)$$

where $q(i)$ is the quantization factor for the coefficient i.

The second model is a Gaussian distribution, and it models the quantization noise as a linear sum of independent noise processes (see Park et al. [19, 20] and Segall et al. [15, 18, 21, 22]). This second distribution is written as

$$P_K(\mathbf{y}_l \mid \mathbf{f}_k, \mathbf{d}_{l,k}) \propto \exp\left[-\frac{1}{2}\big(\mathbf{y}_l - \mathbf{AHC}(\mathbf{d}_{l,k})\mathbf{f}_k\big)^T \mathbf{K}_Q^{-1}\big(\mathbf{y}_l - \mathbf{AHC}(\mathbf{d}_{l,k})\mathbf{f}_k\big)\right], \quad (9.9)$$

where $\mathbf{K}_Q$ is the covariance matrix that describes the noise (see Segall et al. [15] for details).

The LR motion vectors $\mathbf{v}_l$ used to predict the LR images are available after compression; therefore we can also include them in the observation process. We are interested in using the LR motion vectors $\mathbf{v}_{l,k}$ provided by the decoder. Note, however, that not all $\mathbf{y}_l$ are, during the coding process, predicted from frame k.

Although it has been less studied in the literature than the quantization noise distribution, the LR motion vectors are also important for estimating $\mathbf{f}_k$ and $\mathbf{d}$. Various authors have provided different approaches to model $P(\mathbf{v}_{l,k} | \mathbf{f}_k, \mathbf{d}_{l,k}, \mathbf{y}_l)$. Chen and Shultz [23] propose the following distribution

$$P_{CS}(\mathbf{v}_{l,k} | \mathbf{f}_k, \mathbf{d}_{l,k}, \mathbf{y}_l) = \begin{cases} constant \ \ if \ \left| v_{l,k}(j) - \left[\mathbf{A}_D \mathbf{d}_{l,k}\right](j) \right| \leq \Delta, \ \forall j, \\ 0 \ \ elsewhere \end{cases} \quad (9.10)$$

where $\mathbf{A}_D$ is a matrix that maps the displacements to the LR grid, Δ denotes the maximum difference between the transmitted motion vectors and estimated displacements, and $\left[\mathbf{A}_D \mathbf{d}_{l,k}\right](j)$ is the jth element of the vector $\mathbf{A}_D \mathbf{d}_{l,k}$. This distribution enforces the motion vectors to be close to the actual subpixel displacements and represents a reasonable condition.

A similar idea is proposed in Mateos et al. [24] where the following distribution is used

$$P_M(\mathbf{v}_{l,k} | \mathbf{f}_k, \mathbf{d}_{l,k}, \mathbf{y}_l) \propto \left[-\frac{\gamma_l}{2} \left\| \bar{\mathbf{v}}_{l,k} - \mathbf{d}_{l,k} \right\|^2 \right], \quad (9.11)$$

where γ_l specifies the similarity between the transmitted and estimated information and $\bar{\mathbf{v}}_{l,k}$ denotes an upsampled version of $\mathbf{v}_{l,k}$ to the size of the HR motion vectors.

Segall et al. in [21] and [22] model the displaced frame difference within the encoder using the distribution

$$P_K(\mathbf{v}_{l,k} | \mathbf{f}_k, \mathbf{d}_{l,k}, \mathbf{y}_l) \propto \left[-\frac{1}{2} \left(MC_l(\mathbf{y}_l^P, \mathbf{v}_l) - \mathbf{AHC}(\mathbf{d}_{l,k})\mathbf{f}_k \right)^T \mathbf{K}_{MV}^{-1} \left(MC_l(\mathbf{y}_l^P, \mathbf{v}_l) - \mathbf{AHC}(\mathbf{d}_{l,k})\mathbf{f}_k \right) \right],$$

$$(9.12)$$

where $\mathbf{K}_{MV}$ is the covariance matrix for the prediction error between the original frame once in LR $\left(\mathbf{AHC}(\mathbf{d}_{l,k})\mathbf{f}_k \right)$ and its motion compensation estimate $(MC_l(\mathbf{y}_l^P, \mathbf{v}_l))$.

Note that from the observation models of the LR compressed images and motion vectors we have described, we can write the joint observational model of the LR compressed images and LR motion vectors given the HR image and motion vectors as

$$P(\mathbf{y}, \mathbf{v} \mid \mathbf{f}_k, \mathbf{d}) = \prod_l P(\mathbf{y}_{l,k} \mid \mathbf{f}_k, \mathbf{d}_{l,k}) P(\mathbf{v}_{l,k} \mid \mathbf{f}_k, \mathbf{d}_{l,k}, \mathbf{y}_l). \tag{9.13}$$

9.3   Regularization in HR

The super-resolution problem is an ill-posed problem. Given an (compressed or uncompressed) LR sequence, the estimation of the HR image and motion vectors maximizing any of the conditional distributions that describe the acquisition models shown in the previous section is a typical example of an ill-posed problem. Therefore, we have to regularize the solution or, using statistical language, introduce *a priori* models on the HR image and/or motion vectors.

9.3.1   Uncompressed observations

Maximum likelihood (ML), maximum *a posteriori* (MAP), and the set theoretic approach using POCS can be used to provide solutions of the SR problem (see Elad and Feuer [5]).

In all published work on HR, it is assumed that the variables $\mathbf{f}_k$ and $\mathbf{d}$ are independent, that is,

$$P(\mathbf{f}_k, \mathbf{d}) = P(\mathbf{f}_k)P(\mathbf{d}). \tag{9.14}$$

Some HR reconstruction methods give the same probability to all possible HR images $\mathbf{f}_k$ and motion vectors $\mathbf{d}$ (see for instance, Stark and Oskoui [10] and Tekalp et al. [11]). This would also be the case of the work by Irani and Peleg [7] (see also references in [9] for the so-called *simulate* and *correct methods*).

Giving the same probability to all possible HR images $\mathbf{f}_k$ and motion vectors $\mathbf{d}$ is equivalent to using the noninformative prior distributions

$$P(\mathbf{f}_k) \propto constant, \tag{9.15}$$

and

$$P(\mathbf{d}) \propto constant. \tag{9.16}$$

Note that although POCS is the method used in [10, 11] to find the HR image, no prior information is included on the image we try to estimate.

Most of the work on POCS for HR estimation uses the acquisition model that imposes constraints on the maximum difference between each component of $\mathbf{g}_l$ and $\mathbf{AHC}(\mathbf{d}_{l,k})\mathbf{f}_k$. This model corresponds to uniform noise modeling, with no regularization on the HR image. See, however, [9] for the introduction of convex sets as *a priori* constraints on the image using the POCS formulation, and also Tom and Katsaggelos [25].

What do we know *a priori* about the HR images? We expect the images to be smooth within homogeneous regions. A typical choice to model this idea is the following prior distribution for $\mathbf{f}_k$:

$$P(\mathbf{f}_k) \propto \exp\left[-\left(\frac{\lambda_1}{2}\left\|\mathbf{Q}_1\mathbf{f}_k\right\|^2\right)\right], \qquad (9.17)$$

where $\mathbf{Q}_1$ represents a linear high-pass operation that penalizes the estimation that is not smooth and λ_1 controls the variance of the prior distribution (the higher the value of λ_1, the smaller the variance of the distribution).

There are many possible choices for the prior model on the HR original image. Two well-known and used models are the Huber's type, proposed by Schultz and Stevenson [3], and the model proposed by Hardie et al. [4] (see Park et al. [26] for additional references on prior models). More recently, total variation methods [27], anisotropic diffusion [28], and compound models [29] have been applied to super-resolution problems.

9.3.2 *Compressed observations*

Together with smoothness, the additional *a priori* information we can include when dealing with compressed observations is that the HR image should not be affected by coding artifacts.

The distribution for $P(\mathbf{f}_k)$ is very similar to the one shown in Equation (9.17), but now we have an additional term in it that enforces smoothness in the LR image obtained from the HR image using our model. The equation becomes

$$P(\mathbf{f}_k) \propto \exp\left[-\left(\frac{\lambda_3}{2}\left\|\mathbf{Q}_3\mathbf{f}_k\right\|^2 + \frac{\lambda_4}{2}\left\|\mathbf{Q}_4\mathbf{AHf}_k\right\|^2\right)\right], \qquad (9.18)$$

where $\mathbf{Q}_3$ represents a linear high-pass operation that penalizes the estimation that is not smooth, $\mathbf{Q}_4$ represents a linear high-pass operator that penalizes estimates with block boundaries, and λ_3 and λ_4 control the weight of the norms. For a complete study of all the prior models used in this problems, see Segall et al. [15].

In the compressed domain, constraints have also been used on the HR motion vectors. Assuming that the displacements are independent between frames (an assumption that maybe should be reevaluated in the future), we can write

$$P(\mathbf{d}) = \prod_l P(\mathbf{d}_{l,k}). \qquad (9.19)$$

We can enforce $\mathbf{d}_k$ to be smooth within each frame and then write

$$P(\mathbf{d}_{l,k}) \propto \exp\left[-\frac{\lambda_5}{2}\left\|\mathbf{Q}_5\mathbf{d}_{l,k}\right\|^2\right],$$ (9.20)

where $\mathbf{Q}_5$ represents a linear high-pass operation that, once again, penalizes the displacement estimation that is not smooth and λ_5 controls the variance of the distribution (see, again, Segall et al. [15] for details).

9.4 Estimating HR images

Having described in the previous sections the resolution image and motion vector priors and the acquisition models used in the literature, we now turn our attention to computing the HR frame and motion vectors.

Since we have introduced both priors and conditional distributions, we can apply the Bayesian paradigm in order to find the maximum of the posterior distribution of HR image and motion vectors given the observations.

9.4.1 Uncompressed sequences

For uncompressed sequences our goal becomes finding $\hat{\mathbf{f}}_k$ and $\hat{\mathbf{d}}$ that satisfy

$$\hat{\mathbf{f}}_k, \hat{\mathbf{d}} = \arg\max_{\mathbf{f}_k, \mathbf{d}} P(\mathbf{f}_k, \mathbf{d})P(\mathbf{g}\,|\,\mathbf{f}_k, \mathbf{d}),$$ (9.21)

where the prior distributions, $P(\mathbf{f}_k, \mathbf{d})$, used for uncompressed sequences have been defined in Subsection 9.3.1, and the acquisition models for this problem, $P(\mathbf{g}\,|\,\mathbf{f}_k, \mathbf{d})$, are described in Subsection 9.2.1.

Most of the reported works in the literature on SR from uncompressed sequences first estimate the HR motion vectors either by first interpolating the LR observations and then finding the HR motion vectors or by first finding the motion vectors in the LR domain and then interpolating them. Thus, classical motion estimation or image registration techniques can be applied to the process of finding the HR motion vectors (see, for instance, Brown [30], Szeliski [31], and Stiller and Konrad [32]).

Interesting models for motion estimation have also been developed within the HR context. These works also perform segmentation within the HR image (see, for instance, Irani and Peleg [7] and Eren et al. [13]).

Once the HR motion vectors, denoted by $\overline{\mathbf{d}}$, have been estimated, all the methods proposed in the literature proceed to find $\overline{\mathbf{f}}_k$ satisfying

$$\overline{\mathbf{f}}_k = \arg\max_{\mathbf{f}_k} P(\mathbf{f}_k)P(\mathbf{g}\,|\,\mathbf{f}_k, \overline{\mathbf{d}}).$$ (9.22)

Several approaches have been used in the literature to find $\bar{f}_k$, for example, gradient descent, conjugate gradient, preconditioning, and POCS (see Borman and Stevenson [9] and Park et al. [26]).

Before leaving this section, it is important to note that some work has been developed in the uncompressed domain to carry out the estimation of the HR image and motion vectors simultaneously (see Tom and Katsaggelos [33, 34, 35] and Hardie et al. [36]).

9.4.2 Compressed sequences

For compressed sequences our goal becomes finding $\hat{f}_k$ and $\hat{d}$ that satisfy

$$\hat{f}_k, \hat{d} = \arg\max_{f_k, d} P(f_k, d)P(y, v \mid f_k, d), \qquad (9.23)$$

where the distributions of HR intensities and displacements used in the literature have been described in Subsection 9.3.2 and the acquisition models have already been studied in Subsection 9.2.2. The solution is found with a combination of gradient descent, nonlinear projection, and full-search methods. Scenarios where d is already known or separately estimated are a special case of the resulting procedure.

One way to find the solution of Equation (9.23) is by using the cyclic coordinate descent procedure [37]. An estimate for the displacements is first found by assuming that the HR image is known, so that

$$\hat{d}^{q+1} = \arg\max_{d} P(d)P(y, v \mid \hat{f}_k^q, d), \qquad (9.24)$$

where q is the iteration index for the joint estimate. (For the case where d is known, Equation (9.24) becomes $\hat{d}^{q+1} = \bar{d} \, \forall q$.) The intensity information is then estimated by assuming that the displacement estimates are exact, that is

$$\hat{f}_k^{q+1} = \arg\max_{f_k} P(f_k)P(y, v \mid f_k, \hat{d}^{q+1}). \qquad (9.25)$$

The displacement information is reestimated with the result from Equation (9.25), and the process iterates until convergence. The remaining question is how to solve Equations (9.24) and (9.25) for the distributions presented in the previous sections.

The noninformative prior in Equation (9.16) is a common choice for $P(d)$. Block-matching algorithms are well suited for solving Equation (9.24) for this particular case. The construction of $P(y, v \mid f_k, d)$ controls the performance of the block-matching procedure (see Mateos et al. [24], Segall et al. [21, 22], and Chen and Schultz [23]).

When $P(\mathbf{d})$ is not uniform, differential methods become common methods for the estimation of the displacements. These methods are based on the optical flow equation and are explored in Segall et al. [21, 22]. An alternative differential approach is utilized by Park et al. [19, 38]. In these works, the motion between LR frames is estimated with the block-based optical flow method suggested by Lucas and Kanade [39]. Displacements are estimated for the LR frames in this case.

Methods for estimating the HR intensities from Equation (9.25) are largely determined by the acquisition model used. For example, consider the least complicated combination of the quantization constraint in Equation (9.8) with the noninformative distributions for $P(\mathbf{f}_k)$ and $P(\mathbf{v}\,|\,\mathbf{f}_k,\mathbf{d},\mathbf{y})$. Note that the solution to this problem is not unique, as the set-theoretic method only limits the magnitude of the quantization error in the system model. A frame that satisfies the constraint is therefore found with the POCS algorithm [40], in which sources for the projection equations include [17] and [18].

A different approach must be followed when incorporating the spatial domain noise model in Equation (9.9). If we still assume a noninformative distribution for $P(\mathbf{f}_k)$ and $P(\mathbf{v}\,|\,\mathbf{f}_k,\mathbf{d},\mathbf{y})$, the estimate can be found with a gradient descent algorithm [18].

Figure 9.8 shows one example of the use of the techniques just described to estimate an HR frame.

9.5   Parameter estimation in HR

Since the early work by Tsai and Huang [8], researchers, primarily within the engineering community, have focused on formulating the HR problem as a reconstruction or a recognition one. However, as reported in [9], not much work has been devoted to the efficient calculation of the reconstruction or to the estimation of the associated parameters. In this section we will briefly review these two very interesting research areas.

Bose and Boo [41] use a block semi-circulant (BSC) matrix decomposition in order to calculate the MAP reconstruction; Chan et al. [42, 43, 44] and Nguyen [6, 45, 46,] use preconditioning, wavelets, as well as BSC matrix decomposition. The efficient calculation of the MAP reconstruction is also addressed by Ng et al. [47, 48] and Elad and Hel-Or [49].

To our knowledge, only the works by Bose et al. [50], Nguyen [6, 46, 51, 52], and to some extent [34, 44, 53, 54] address the problem of parameter estimation. Furthermore, in those works the same parameter is assumed for all the LR images, although in the case of [50] the proposed method can be extended to different parameter for LR images (see [55]).

Recently, Molina et al. [56] have used the general framework for frequency domain multichannel signal processing developed by Katsaggelos et al. in [57] and Banham et al. in [58] (a formulation that was also obtained later by Bose and Boo [41] for the HR problem) to tackle the parameter estimation in HR problems. With the use of BSC matrices, the authors show that all the matrix calculations involved in the parameter maximum likeli-

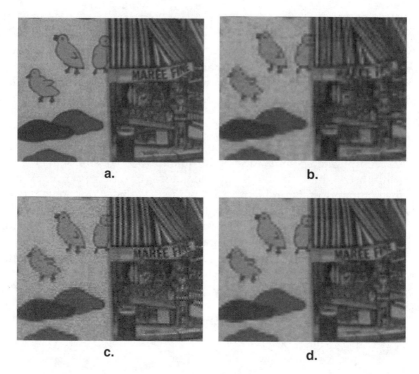

Figure 9.8 From a video sequence: (a) original image; (b) decoded result after bilinear interpolation. The original image is downsampled by a factor of two in both the horizontal and vertical directions and then compressed with an MPEG-4 encoder operation at 1 Mb/s; (c) super-resolution image employing only the quantization constraint in Equation (9.8); (d) super-resolution image employing the normal approximation for $P(\mathbf{y} \mid \mathbf{f}_k, \mathbf{d})$ in Equation (9.9), the distribution of the LR motion vectors given by Equation (9.12), and the HR image prior in Equation (9.18).

hood estimation can be performed in the Fourier domain. The proposed approach can be used to assign the same parameter to all LR images or make them image dependent. The role played by the number of available LR images in both the estimation procedure and the quality of the reconstruction is examined in [59] and [60].

Figure 9.9(a) shows the upsampled version, 128×64, of one 32×16 LR observed image. We ran the reconstruction algorithm in [60] using 1, 2, 4, 8, and 16 LR subpixel shifted images.

Before leaving this section, we would like to mention that the above reviewed works address the problem of parameter estimation for the case of multiple undersampled, shifted, degraded frames with subpixel displacement errors and that very interesting areas to be explored are the cases of general compressed or uncompressed sequences.

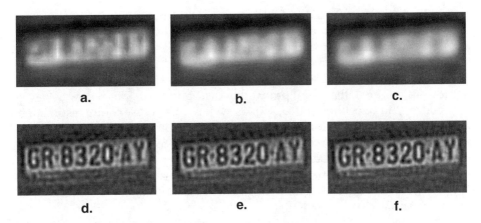

Figure 9.9 Spanish car license plate example: (a) unsampled observed LR image; (b)–(f) reconstruction using 1, 2, 4, 8, and 16 LR images.

9.6 New approaches toward HR

To conclude this chapter, we want to identify several research areas that we believe will benefit the field of super-resolution from video sequences.

A key area is the simultaneous estimate of multiple HR frames. These sequence estimates can incorporate additional spatio-temporal descriptions for the sequence and provide increased flexibility in modeling the scene. For example, the temporal evolution of the displacements can be modeled. Note that there is already some work in both compressed and uncompressed domains (see Hong et al. [61, 62], Choi et al. [63], and Dekeyser et al. [64]).

Alvarez et al. [65] have also addressed the problem of simultaneous reconstruction of compressed video sequences. To encapsulate the statement that the HR images are correlated and absent of blocking artifacts, the prior distribution

$$P(\mathbf{f}_1, \ldots, \mathbf{f}_L) \propto \exp\left\{ -\frac{\lambda_1}{2} \sum_{l=2}^{L} \left\| \mathbf{f}_{l-1} - \mathbf{C}(\mathbf{d}_{l-1,l}) \mathbf{f}_l \right\|^2 - \frac{\lambda_2}{2} \sum_{l=1}^{L} \left\| \mathbf{Q}_2 \mathbf{f}_l \right\|^2 - \frac{\lambda_3}{2} \sum_{l=1}^{L} \left\| \mathbf{Q}_3 \mathbf{A} \mathbf{H} \mathbf{f}_l \right\|^2 \right\}$$

(9.26)

is utilized. Here, $\mathbf{Q}_2$ represents a linear high-pass operation that penalizes super-resolution estimates that are not smooth, $\mathbf{Q}_3$ represents a linear high-pass operator that penalizes estimates with block boundaries, and λ_1, λ_2, and λ_3 control the weight given to each term. A common choice for $\mathbf{Q}_2$ is the discrete 2-D Laplacian; a common choice for $\mathbf{Q}_3$ is the simple difference operation applied at the boundary locations. The HR motion vectors are previously estimated. Figure 9.10 shows just one image obtained by the method proposed in [65] together with the bilinear interpolation of the compressed LR observation.

Accurate estimates of the HR displacements are critical for the super-resolution problem. There is work to be done in designing methods for the blurred, subsampled, aliased, and, in some cases, blocky observations. Toward this goal, the use of probability distribution of optical flow as developed by Simoncelli et al. [66, 67], as well as the coarse-to-fine estimation, seem areas worth exploring. Note that we could incorporate also the coherence of the distribution of the optical flow when constructing the super-resolution image (or sequence).

The use of band-pass directional filters on super-resolution problems also seems an interesting area of research (see Nestares and Navarro [68] and Chamorro-Martínez [69]).

Let us consider the digits in a car plate shown in Figure 9.11(a). The digits after blurring, subsampling, and noise are almost undistinguishable, as shown in Figure 9.11(b). Following Baker and Kanade [70], there are limits for the approach discussed in Molina et al. [56] (see Figure 9.9). However, any classification/recognition method will benefit from resolution improvement on the observed images. Would it be possible to learn the prior image model from a training set of HR images that have undergone a blurring, subsampling, and, in some cases, compression process to produce images similar to the observed LR ones? This approach has been pursued by several authors, and it will be described now in some depth.

Baker and Kanade [70] approach the super-resolution problem in restricted domains (see the face recognition problem in [70]) by trying to learn the prior model from a set of training images. Let us assume that we have a set of HR training images, the size of each image being $2^N \times 2^N$. For all the training images, we can form their feature pyramids. These pyramids of features are created by calculating, at different resolutions, the Laplacian L, the horizontal H and vertical V first derivatives, and the horizontal H_2 and

a.   b.

Figure 9.10 (a) Bilinear interpolation of an LR compressed image; (b) image obtained when the whole HR sequence is processed simultaneously.

a.   **b.**

Figure 9.11 (a) HR image; (b) corresponding LR observation.

vertical V_2 second derivatives. Thus, for a given high training HR image I, we have at resolution $j, j = 0, \ldots, N$ (the higher the value of j, the smaller the resolution), the data

$$F_j(I) = (L_j(I), H_j(I), V_j(I), H_j^2(I), V_j^2(I)). \tag{9.27}$$

We are now given an LR $2^K \times 2^K$ image Lo, with $K < N$, which is assumed to have been registered (motion compensated) with regard to the position of the HR images in the training set. Obviously, we can also calculate for $j = 0, \ldots, K$

$$F_j(Lo) = (L_j(Lo), H_j(Lo), V_j(Lo), H_j^2(Lo), V_j^2(Lo)). \tag{9.28}$$

Let us assume that we have now a pixel (x,y) on the LR observed image and we want to improve its resolution. We can now find the HR image in the training set whose LR image of size $2^K \times 2^K$ has the pixel on its position (m_x,m_y) with the most similar pyramid structure to the one built for pixel (x,y). Obviously, there are different ways to define the concept of more similar pyramids of features (see [70]).

   Now we note that the LR pixel (m_x,m_y) comes from $2^{N-K} \times 2^{N-K}$ HR pixels. On each of these HR pixels we calculate the horizontal and vertical derivatives, and as HR image prior, we force the similarity between these derivatives and the one calculated in the corresponding position of the HR image we want to find.

   As an acquisition model, the authors use the one given by Equation (9.5). Thus, several LR observations can be used to calculate the HR images although, so far, only one of them is used to build the prior. Then for this combination of learned HR image prior and acquisition model, the authors estimate the HR image.

   Instead of using the training set to define the derivative-based prior model, Capel and Zisserman [71] use it to create a principal component basis and then express the HR image we are looking for as a linear combination of the elements of a subspace of this basis. Thus, our problem has now become the calculation of the coefficient of the vectors in the subspace.

Priors can then be defined either on those coefficients or in terms of the real HR underlying image (see [71] for details). Similar ideas have later been developed by Gunturk et al. [72].

It is also important to note that, again, for this kind of priors, several LR images, registered or motion compensated, can be used in the degradation model given by Equation (9.5).

The third alternative approach to super-resolution from video sequences we will consider here is the one developed by Bishop et al. [73]. This approach builds on the one proposed by Freeman et al. [74] for the case of static images.

The approach proposed in [74] involves the assembly of a large database of patch pairs. Each pair is made of an HR patch, usually of size 5×5, and a corresponding patch of size 7×7. This corresponding patch is built as follows: The HR image, I, is first blurred and downsampled (following the process to obtain an LR image from an HR one), and then this LR image is upsampled to have the same size as the HR one, which we call I_{ud}. Then, to each 5×5 patch in I, we associate the 7×7 patch in I_{ud} centered in the same pixel position as the 5×5 HR patch. To both the HR and the upsampled LR images, low frequencies are removed, so it could be assumed that we have the high frequencies in the HR image and the mid-frequencies in the upsampled LR one.

From those 5×5 and 7×7 patches, we can build a dictionary; the patches are also contrast normalized to avoid having a huge dictionary. We now want to create the HR image corresponding to an LR one. The LR image is upsampled to the size of the HR image. Given a location of a 5×5 patch that we want to create, we use its corresponding 7×7 upsampled LR image to search the dictionary and find its corresponding HR image. In order to enforce consistency of the 5×5 patches, they are overlapped (see Freeman et al. [74]). Two alternative methods for finding the best HR image estimate are proposed in [75].

Bishop et al. [73] modify the cost function to be minimized when finding the most similar 5×5 patch by including terms that penalize flickering and that enforce temporal coherence.

Note that from the above discussion on the work of Freeman et al. [74], work on vector quantization can also be extended and applied to the super-resolution problem (see Nakagaki and Katsaggelos [76]).

References

[1]   S. Chaudhuri (Ed.), *Super-Resolution from Compressed Video*, Kluwer Academic Publishers, Boston, 2001.

[2]   M.G. Kang and S. Chaudhuri (Eds.), Super-resolution image reconstruction, *IEEE Signal Processing Magazine*, vol. 20, no. 3, 2003.

[3]   R.R. Schultz and R.L. Stevenson, Extraction of high resolution frames from video sequences, *IEEE Transactions on Image Processing*, vol. 5, pp. 996–1011, 1996.

[4] R.C. Hardie, K.J. Barnard, J.G. Bognar, E.E. Armstrong, and E.A. Watson, High resolution image reconstruction from a sequence of rotated and translated frames and its application to an infrared imaging system, *Optical Engineering*, vol. 73, pp. 247–260, 1998.

[5] M. Elad and A. Feuer, Restoration of a single superresolution image from several blurred, noisy, and undersampled measured images, *IEEE Transactions on Image Processing*, vol. 6, pp. 1646–1658, 1997.

[6] N. Nguyen, P. Milanfar, and G. Golub, A computationally efficient superresolution image reconstruction algorithm, *IEEE Transactions on Image Processing*, vol. 10, pp. 573–583, 2001.

[7] M. Irani and S. Peleg, Motion analysis for image enhancement: resolution, occlusion, and transparency, *Journal of Visual Communication and Image Representation*, vol. 4, pp. 324–335, 1993.

[8] R.Y. Tsai and T.S. Huang, Multiframe image restoration and registration, *Advances in Computer Vision and Image Processing*, vol. 1, pp. 317–339, 1984.

[9] S. Borman and R. Stevenson, Spatial resolution enhancement of low-resolution image sequences: a comprehensive review with directions for future research, Technical report, Laboratory for Image and Signal Analysis (LISA), University of Notre Dame, July 1998.

[10] H. Stark and P. Oskoui, High resolution image recovery from image-plane arrays, using convex projections, *Journal of the Optical Society of America A*, vol. 6, pp. 1715–1726, 1989.

[11] A.M. Tekalp, M.K. Ozkan, and M.I. Sezan, High-resolution image reconstruction from lower-resolution image sequences and space varying image restoration, *Proceedings of the IEEE International Conference on Acoustics, Speech, and Signal Processing*, pp. 169–172, 1992.

[12] A.J. Patti, M.I. Sezan, and A.M. Tekalp, Superresolution video reconstruction with arbitrary sampling lattices and nonzero aperture time, *IEEE Transactions on Image Processing*, vol. 6, pp. 1064–1076, 1997.

[13] P.E. Eren, M.I. Sezan, and A.M. Tekalp, Robust, object-based high-resolution image reconstruction from low-resolution video, *IEEE Transactions on Image Processing*, vol. 6, pp. 1446–1451, 1997.

[14] A.M. Tekalp, *Digital Video Processing*, Signal Processing Series, Prentice Hall, Englewood Cliffs, NJ, 1995.

[15] C.A. Segall, R. Molina, and A.K. Katsaggelos, High-resolution images from low-resolution compressed video, *IEEE Signal Processing Magazine*, vol. 20, pp. 37–48, 2003.

[16] Y. Altunbasak, A.J. Patti, and R.M. Mersereau, Super-resolution still and video reconstruction from MPEG-coded video, *IEEE Transactions on Circuits and Systems for Video Technology*, vol. 12, pp. 217–226, 2002.

[17] A.J. Patti and Y. Altunbasak, Super-resolution image estimation for transform coded video with application to MPEG, *Proceedings of the IEEE International Conference on Image Processing*, vol. 3, pp. 179–183, 1999.

[18] C.A. Segall, A.K. Katsaggelos, R. Molina, and J. Mateos, Super-resolution from compressed video, in *Super-Resolution Imaging*, S. Chaudhuri, Ed., pp. 211–242, Kluwer Academic Publishers, Boston, 1992.

[19] S.C. Park, M.G. Kang, C.A. Segall, and A.K. Katsaggelos, High-resolution image reconstruction of low-resolution DCT-based compressed images, *IEEE International Conference on Acoustics, Speech, and Signal Processing*, vol. 2, pp. 1665–1668, 2002.

[20] S.C. Park, M.G. Kang, C.A. Segall, and A.K. Katsaggelos, Spatially adaptive high-resolution image reconstruction of low-resolution DCT-based compressed images, *IEEE Transactions on Image Processing*, vol. 13, pp. 573–585, 2004.

[21] C.A. Segall, R. Molina, A.K. Katsaggelos, and J. Mateos, Reconstruction of high-resolution image frames from a sequence of low-resolution and compressed observations, *Proceedings of the IEEE International Conference on Acoustics, Speech, and Signal Processing*, vol. 2, pp. 1701–1704, 2002.

[22] C.A. Segall, R. Molina, A.K. Katsaggelos, and J. Mateos, Bayesian resolution enhancement of compressed video, *IEEE Transactions on Image Processing*, forthcoming 2004.

[23] D. Chen and R.R. Schultz, Extraction of high-resolution video stills from MPEG image sequences, *Proceedings of the IEEE Conference on Image Processing*, vol. 2, pp. 465–469, 1998.

[24] J. Mateos, A.K. Katsaggelos, and R. Molina, Simultaneous motion estimation and resolution enhancement of compressed low resolution video, *IEEE International Conference on Image Processing*, vol. 2, pp. 653–656, 2000.

[25] B.C. Tom and A.K. Katsaggelos, An iterative algorithm for improving the resolution of video sequences, *Proceedings of SPIE Conference on Visual Communications and Image Processing*, vol. 2727, pp. 1430–1438, 1996.

[26] S.C. Park, M.K. Park, and M.G. Kang, Super-resolution image reconstruction: a technical overview, *IEEE Signal Processing Magazine*, vol. 20, pp. 21–36, 2003.

[27] D.P. Capel and A. Zisserman, Super-resolution enhancement of text image sequence, *International Conference on Pattern Recognition*, vol. 6, pp. 600–605, 2000.

[28] H. Kim, J.-H. Jang, and K.-S. Hong, Edge-enhancing super-resolution using anistropic diffusion, *Proceedings of the IEEE Conference on Image Processing*, vol. 3, pp. 130–133, 2001.

[29] D. Rajan and S. Chaudhuri, Generation of super-resolution images from blurred observations using an MRF model, *Journal of Mathematical Imaging Vision*, vol. 16, pp. 5–15, 2002.

[30] L.G. Brown, A survey of image registration techniques, *ACM Computing Surveys*, vol. 24, pp. 325–376, 1992.

[31] R. Szeliski, Spline-based image registration, *International Journal of Computer Vision*, vol. 22, pp. 199–218, 1997.

[32] C. Stiller and J. Konrad, Estimating motion in image sequences, *IEEE Signal Processing Magazine*, vol. 16, pp. 70–91, 1999.

[33] B.C. Tom and A.K. Katsaggelos, Reconstruction of a high-resolution image from multiple-degraded and misregistered low-resolution images, *Proceedings of SPIE Conference on Visual Communication and Image Processing*, vol. 2308, pp. 971–981, 1994.

[34] B.C. Tom and A.K. Katsaggelos, Reconstruction of a high-resolution image by simultaneous registration, restoration, and interpolation of low-resolution images, *Proceedings of the IEEE International Conference on Image Processing*, vol. 2, pp. 539–542, 1995.

[35] B.C. Tom and A.K. Katsaggelos, Resolution enhancement of monochrome and color video using motion compensation, *IEEE Transactions on Image Processing*, vol. 10, pp. 278–287, 2001.

[36] R.C. Hardie, K.J. Barnard, and E.E. Armstrong, Joint map registration and high-resolution image estimation using a sequence of undersampled images, *IEEE Transactions on Image Processing,* vol. 6, pp. 1621–1633, 1997.

[37] D.G. Luenberger, *Linear and Nonlinear Programming,* Addison-Wesley, Reading, MA, 1984.

[38] S.C. Park, M.G. Kang, C.A. Segall, and A.K. Katsaggelos, Spatially adaptive high-resolution image reconstruction of low-resolution DCT-based compressed images, *Proceedings of the IEEE Conference on Image Processing,* vol. 2, pp. 861–864, 2002.

[39] B.D. Lucas and T. Kanade, An iterative image registration technique with an application to stereo vision, *Proceedings of the Image Understanding Workshop,* pp. 121–130, 1981.

[40] D.C. Youla and H. Webb, Image restoration by the method of convex projections: part 1 — theory, *IEEE Transactions on Medical the Image,* vol. MI-1, no. 2, pp. 81–94, 1982.

[41] N.K. Bose and K.J. Boo, High-resolution image reconstruction with multisensors, *International Journal of Imaging Systems and Technology,* vol. 9, pp. 141–163, 1998.

[42] R.H. Chan, T.F. Chan, M.K. Ng, W.C. Tang, and C.K. Wong, Preconditioned iterative methods for high-resolution image reconstruction with multisensors, *Proceedings to the SPIE Symposium on Advanced Signal Processing: Algorithms, Architectures, and Implementations,* vol. 3461, pp. 348–357, 1998.

[43] R.H. Chan, T.F. Chan, L. Shen, and S. Zuowei, A wavelet method for high-resolution image reconstruction with displacement errors, *Proceedings of the International Symposium on Intelligent, Multimedia, Video, and Speech Processing,* pp. 24–27, 2001.

[44] R.H. Chan, T.F. Chan, L.X. Shen, and Z.W. Shen, Wavelet algorithms for high-resolution image reconstruction, Technical report, Department of Mathematics, Chinese University of Hong Kong, 2001.

[45] N. Nguyen and P. Milanfar, A wavelet-based interpolation–restoration method for super-resolution, *Circuits, Systems, and Signal Processing,* vol. 19, pp. 321–338, 2000.

[46] N. Nguyen, *Numerical Algorithms for Superresolution,* Ph.D. thesis, Stanford University, 2001.

[47] M.K. Ng, R.H. Chan, and T.F. Chan, Cosine transform preconditioners for high-resolution image reconstruction, *Linear Algebra and Its Applications,* vol. 316, pp. 89–104, 2000.

[48] M.K. Ng and A.M. Yip, A fast MAP algorithm for high-resolution image reconstruction with multisensors, *Multidimensional Systems and Signal Processing,* vol. 12, pp. 143–164, 2001.

[49] M. Elad and Y. Hel-Or, A fast super-resolution reconstruction algorithm for pure translational motion and common space invariant blur, *IEEE Transactions on Image Processing,* vol. 10, pp. 1187–1193, 2001.

[50] N.K. Bose, S. Lertrattanapanich, and J. Koo, Advances in superresolution using L-curve, *IEEE International Symposium on Circuits and Systems,* vol. 2, pp. 433–436, 2001.

[51] N. Nguyen, P. Milanfar, and G. Golub, Blind superresolution with generalized cross-validation using Gauss-type quadrature rules, *33rd Asilomar Conference on Signal, Systems, and Computers,* vol. 2, pp. 1257–1261, 1999.

[52] N. Nguyen, P. Milanfar, and G. Golub, Efficient generalized cross-validation with applications to parametric image restoration and resolution enhancement, *IEEE Transactions on Image Processing*, vol 10, pp. 1299–1308, 2001.

[53] B.C. Tom, N.P. Galatsanos, and A.K. Katsaggelos, Reconstruction of a high resolution image from multiple low resolution images, in *Super-Resolution Imaging*, S. Chaudhuri, Ed., pp. 73–105, Kluwer Academic Publishers, Boston, 1992.

[54] B.C Tom, A.K. Katsaggelos, and N.P. Galatsanos, Reconstruction of a high-resolution image from registration and restoration of low-resolution images, *Proceedings of the IEEE International Conference on Image Processing*, vol. 3, pp. 553–557, 1994.

[55] M. Belge, M.E. Kilmer, and E.L. Miller, Simultaneous multiple regularization parameter selection by means of the L-hypersurface with applications to linear inverse problems posed in the wavelet domain, *Proceedings of SPIE'98: Bayesian Inference Inverse Problems*, 1998.

[56] R. Molina, M. Vega, J. Abad, and A.K. Katsaggelos, Parameter estimation in Bayesian high-resolution image reconstruction with multisensors, *IEEE Transactions on Image Processing*, vol. 12, pp. 1655–1667, 2003.

[57] A.K. Katsaggelos, K.T. Lay, and N.P. Galatsanos, A general framework for frequency domain mulitchannel signal processing, *IEEE Transaction on Image Processing*, vol. 2, pp. 417–420, 1993.

[58] M.R. Banham, N.P. Galatsanoa, H.L. Gonzalez, and A.K. Katsaggelos, Multichannel restoration of single channel images using a wavelet-based subband decomposition, *IEEE Transactions on Image Processing*, vol. 3, pp. 821–833, 1994.

[59] M. Vega, J. Mateos, R. Molina, and A.K. Katsaggelos, Bayesian parameter estimation in image reconstruction from subsampled blurred observations, *Proceedings of the IEEE International Conference on Image Processing*, vol. 2, pp. 969–972, 2003.

[60] F.J. Cortijo, S. Villena, R. Molina, and A.K. Katsaggelos, Bayesian superresolution of text image sequences from low-resolution observations, *IEEE Seventh International Symposium on Signal Processing and Its Applications*, vol. I, pp. 421–424, 2003.

[61] M.-C. Hong, M.G. Kang, and A.K. Katsaggelos, An iterative weighted regularized algorithm for improving the resolution of video sequences, *Proceedings of the IEEE International Conference on Image Processing*, vol. II, pp. 474–477, 1997.

[62] M.-C. Hong, M.G. Kang, and A.K. Katsaggelos, Regularized multichannel restoration approach for globally optimal high resolution video, *Proceedings of the SPIE Conference on Visual Communcations and Image Processing*, pp. 1306–1316, 1997.

[63] M.C. Choi, Y. Yang, and N.P. Galatsanos, Multichannel regularized recovery of compressed video sequences, *IEEE Transactions on Circuits and Systems II*, vol. 48, pp. 376–387, 2001.

[64] F. Dekeyser, P. Bouthemy, and P. Pérez, A new algorithm for super-resolution from image sequences, *9th International Conference on Computer Analysis of Images and Patterns, CAIP 2001, Springer-Verlag Lecture Notes in Computer Science 2124*, pp. 473–481, 2001.

[65] L.D. Alvarez, R. Molina, and A.K. Katsaggelos, Mulitchannel reconstruction of video sequences from low-resolution and compressed observations, *8th Iberoamerican Congress on Pattern Recognition, CIARP 2003, Springer-Verlag Lectures Notes in Computer Science 2905*, pp. 46–53, 2003.

[66] E.P. Simoncelli, E.H. Adelson, and D.J. Heeger, Probability distributions of optical flow, *Proceedings of the IEEE Computer Society Conference on Computer Vision and Pattern Recognition*, pp. 310–315, 1991.

[67] E.P. Simoncelli, Bayesian multi-scale differential optical flow, in *Handbook of Computer Vision and Applications*, B. Jähne, H. Haussecker, and P. Geissler (Eds.), Academic Press, New York, 1999.

[68] O. Nestares and R. Navarro, Probabilistic estimation of optical flow in multiple band-pass directional channels, *Image and Vision Computing*, vol. 19, pp. 339–351, 2001.

[69] J. Chamorro-Martínez, *Desarrollo de modelos computacionales de representación de secuencias de imágenes y su aplicación a la estimación de movimiento* (in Spanish), Ph.D. thesis, University of Granada, 2001.

[70] S. Baker and T. Kanade, Limits on super-resolution and how to break them, *IEEE Transactions on Pattern Analysis and Machine Intelligence*, vol. 24, pp. 1167–1183, 2002.

[71] D.P. Capel and A. Zisserman, Super-resolution from multiple views using learnt image models, *Proceedings of the IEEE Computer Society Conference on Computer Vision and Pattern Recognition*, vol. 2, pp. 627–634, 2001.

[72] B.K. Gunturk, A.U. Batur, Y. Altunbasak, M.H. Hayes, and R.M. Mersereau, Eigenface-domain super-resolution for face recognition, *IEEE Transactions on Image Processing*, vol. 12, pp. 597–606, 2003.

[73] C.M. Bishop, A. Blake, and B. Marthi, Super-resolution enhancement of video, in *Proceedings of Artificial Intelligence and Statistics*, M. Bishop and B. Frey (Eds.), Society for Artificial Intelligence and Statistics, 2003.

[74] W.T. Freeman, T.R. Jones, and E.C. Pasztor, Example based super-resolution, *IEEE Computer Graphics and Applications*, vol. 22, pp. 56–65, 2002.

[75] W.T. Freeman, E.C. Pasztor, and O.T. Carmichael, Learning low-level vision, *International Journal of Computer Vision*, vol. 40, pp. 25–47, 2000.

[76] R. Nakagaki and A.K. Katsaggelos, A VQ-based blind image restoration algorithm, *IEEE Transactions on Image Processing*, vol. 12, pp. 1044–1053, 2003.

[77] J. Mateos, A.K. Katsaggelos, and R. Molina, Resolution enhancement of compressed low resolution video, *IEEE International Conference on Acoustics, Speech, and Signal Processing*, vol. 4, pp. 1919–1922, 2000.

Index

A

Abstracts, video summaries, 224
Active contour methods, 19–20,
    see also specific type
Adelson, Burt and, studies, 10
Adelson, Wang and, studies, 11
Affine displacements, 127–131
Affine motion, 116–121
Agnihotri studies, 221
Aguiar studies, 3, 5–72
Akutsu and Tonomura, Taniguchi,
    studies, 224
Alignment, motion computation, 97–100
Al-Mualla studies, 163
Altunbasak studies, 243
Alvarez studies, 4, 233–259
Amir studies, 219
Aperture effect, 81
Archives, *see* Low-resolution sources;
    Restoration; Video
    summarization
Arithmetic coding, errors, 153
ARQ, *see* Retransmission protocols
    (ARQ)
Automated text summarization,
    226–227
Automatic speech recognition, 223

B

Background mosaics generation, 35–41
Baker and Kanade studies, 252
Ballester and Bertalmio studies, 188
Balloon snake approach, 22–23, 31
Band-pass directional filters, 252

Banham studies, 249
Bayesian techniques
    estimating HR images, 247
    missing data reconstruction, 188
    motion computation, 91–92
    sequence restoration, 189–207
    statistical concealment approach, 162
Bender, Teodosio and, studies, 10
BER, *see* Bit error rate (BER)
Bertalmio, Ballester and, studies, 188
Bessel function, 113
B-frames, 165
Bishop studies, 254
Bit error rate (BER), 141, 154
Blind operating system, 142
Block-based motion encoding, 90, 93
Block interleaving, 170
Blotches, 183, 199–201, 204–205
Bluetooth technology, 135
BMA, *see* Boundary matching algorithm
    (BMA)
BMOVIES system, 225
Boo, Bose and, studies, 249
Boon, Wooi, studies, 205
Borman and Stevenson studies, 248
Bornard studies, 188, 207–208
Bose and Boo studies, 249
Bose studies, 249
Boundary matching algorithm (BMA),
    166
BRAVA consortium, 179
Brown studies, 247
Burt and Adelson studies, 10

261